CW01335488

Always

a

Challenge

Tom Kerr 02/03.

Always
a
Challenge

by

Tom Kerr C.B.

An RAE Scientist in the Cold War Years-
A First Hand Account

Copyright 2002 Thomas H. Kerr

First published in the UK by T.H.Kerr C.B.
"Bundu" 13 Kingsley Ave, Camberley. GU15 2NA.

The information in this book is true and complete to the best knowledge of the author. All observations and recommendations represent his own views and not necessarily those of the MoD, other organisations or companies for whom he has worked. All liability is disclaimed in connection with information contained in this book.

Photographs are mostly MoD copyright and the author is grateful for the permission of the Evening Standard, the Aldershot News and the Surrey Herald Newspapers to use some of their material.

ISBN 0-9543436-0-3

Printed in England by Barnwell's Print Ltd, Aylsham, Norfolk. NR11 6ET

Copies can be obtained in the U.K. by sending £25/ copy to the publisher above.

Foreword

by
Sir John Charnley CB

Tom Kerr was recruited to the Government Scientific Service in September 1949 with the rank of Scientific Officer at the Royal Aircraft Establishment, Farnborough (RAE). There followed a remarkable career spanning an extremely wide range of challenging experiences over almost 50 years in which phenomenal advances in aeronautical achievements took place, with dramatic changes in defence policy, economic performance and the aerospace industry. In all his appointments he was closely involved in one or more of these changes. In June 1984 he retired from Government Service having spent the last four years as Director of that same Establishment where a distinguished career began almost 35 years previously. Never taking kindly to inactivity he continued to work relentlessly for another ten years with senior appointments in the aerospace industry.

I first met Tom and his lovely wife Evelyn on their arrival in Farnborough and as our paths crossed on many occasions since, both professionally and socially, it gives me great pleasure to introduce his autobiography.

What began as a straightforward story for family consumption, of a highly successful career spent largely in defence science with the various prestigious, enjoyable associations that accompany such success, has been transformed into a revealing, absorbing account of his many personal achievements in very different technical fields, and of the intricate processes and procedures involved in the procurement of defence equipment. The tangled web of defence policy, resources, military operations and requirements, technical potential and industrial capability is a 'challenging' theme recurring on many pages. From 'within' the system there is an insight at different functional levels and with different management responsibilities, and from 'without', in later years, there is a strongly expressed view as a director of a major industrial contractor, heavily dependent upon defence projects for its continued success or even existence.

The first chapter sets the scene with a warm description of a happy, supportive home and family background, with successes at school, in the RAF as a pilot, and at Durham University. A scientific career in aviation is clearly in prospect with 'above average' characteristics of intellectual ability, communication skills, and leadership qualities already firmly established.

His first real job in the Aerodynamics Department of RAE at age 25 led to an exciting eleven years with responsibility for research and innovative flight experiments on a wide variety of aircraft at low speeds and transonic speeds, initially at Farnborough and later at RAE Bedford. He made an outstanding personal contribution, not without risk, to the understanding of the factors determining the spinning behaviour and spin recovery of unusual configurations. This chapter includes a brief 'essay' on spinning, which will be of particular interest to the more technically erudite. There are similar 'essays' on relevant topics in later chapters.

There followed a twelve-year mid-career period of important operational analysis studies and weapon system assessments at Bomber Command, the Defence Operational Analysis Establishment and on return to RAE. Although the purpose of the studies varied widely the need for a highly structured approach, and the analytical techniques and methods used were related, and required a disciplined, well-organised mind. Tom's grasp was outstanding and he became well known, for his rigorous examination of complex multi-dimensional problems. A tenacious, well-presented defence of his strict logic was sometimes needed to persuade a critical audience to accept controversial and often unwelcome conclusions. 'Essays' on the Cuban crisis, deterrence, European Air Defence and novel Airfield Attack Weapons refer to important studies for which he was responsible.

Against a background of national economic problems, the early seventies saw the growing need for a reduction in the resources devoted to defence research and development and Tom's extensive analytical expertise was harnessed in a move to MOD Headquarters to make proposals for the rationalisation of the R & D Establishments, airfields and Air Fleet. This was not a task he enjoyed, and his appointment in 1974 to be Director of the National Gas Turbine Establishment (NGTE) was most welcome, despite having to implement some of his own proposals.

Six years running NGTE followed by four years as Director of RAE, two exceptional appointments, presented Tom with a long sought after ambition and the opportunity to display management skills of a high order. Pressure on resources was still high and rationalisation of effort and facilities a top priority. After a successful merger of NGTE into RAE, he was faced with major decisions on the future of the Farnborough airfield. He argued strongly in favour of the retention of a General Aviation Airfield with numerous presentations to staff, Trade Unions, the local community and developers. To his great credit, the airfield is being extended commercially on that premise today.

Intense frustration with MOD is the prominent feature of Tom's final eight years in industry with Hunting Engineering. Frequent changes in service requirements, financial constraints and collaboration complications (some international), are high among the problems he faced when the company was bidding for major weapon contracts. In this context, 'essays' on LAW, SR(A)1236, and SR(A)1238 are worth detailed examination.

This book is a 'big' book written by a 'big' man. He welcomes challenges. To some extent he creates them with his own toughness, then attacks them with an optimistic determination and never failing confidence. He abhors and loses patience with loose logic. He radiates an infectious humour with amusing anecdotes, and whenever possible seeks to defuse serious confrontations. I look forward to sampling the Tournados Kerr, if I am able to afford it!

I have thoroughly enjoyed the book for its strong sense of personal achievement and the new light it sheds on many serious topics. I hope you do too.

Contents

4) 1960-1964   Operational Research-
Scientific Adviser to C in C Bomber Command.

The THOR Medium Range Ballistic Missile System.
Readiness
The Cuban Crisis.
Bomber Operations.
Intelligence.
Countermeasures.
H2S. The Navigation and Bombing System.
Blue Steel
Skybolt
Penetration Studies and Conversion to Low Level Operations.
Aircraft Servicing Trials.

5) 1964-1966   Mathematical Assessment Division at Farnborough.

BL755
The General Approach to the Work
The Buccaneer 2*
SAGW-Fleet Defence
The TSR 2. Its Cancellation and subsequent Events.
Fuse/Warhead Matching
Reliability.

6) 1966-1970   Operational   Analysis at D.O.A.E.

The SAM/Fighter Working Party
Studies of Future (post 1975) Air Operations in Europe.
The Maritime Protection Task.
Exercise HELLTANK
Study of Maritime Operations in the 1980's with
Special Reference to Anti-Ship Weapons.
House of Commons Select Committee on Science and
Technology

7) 1970-1972   Head of Weapons Research. RAE.

JP233
The Gulf War
West Freugh

8) 1972-1974   M.O.D. London.

Research Establishment Rationalisation
The Airfields Study
Project Information
The Procurement Executive Air Fleet
Aerodynamics Department NPL Teddington
Naval Engineering
Director Trials and Guided Weapons and the Joint
Other Tasks.
My Next Move.

9) 1974-1980   N.G.T.E. Pyestock.

The R & D Programme.
The Engine Test Facilities
The Naval Marine Wing.
Management and Industrial Relations.
Testing of the FJR710 from Japan.
The Chinese at Pyestock.
Rationalisation.
The Noise Test Facility and the Visit of The Duke Of Edinburgh.
The Apprentice School
ISM Awards
Whittle and the "Reactionaries."
Health and Safety at Work
The Social Life of the Establishment.

10) 1980-1984   R.A.E. Farnborough.

The Reorganisation of RAE.
The Airfield
Contractor Operation of the Ranges.
Contractor Operation of the Wind Tunnels.
The Concorde Structural Test Facility
Integration of NGTE into RAE

The Technology
Computational Fluid Dynamics
Flight Systems
Civil Air Navigation and Colour Cockpit Displays
Fly-by-Wire and Flight Simulation
Space
Materials and Structures Department.
Weapons Department
The Falklands

The Fun

The RAE Apprentices
ISM Awards
75th Anniversary of Cody's First Flight 16th Oct 1908.
The Cody School
The De Havilland DH100 Centenary Rally
The Comet DH 88 Refurbishment
The Passing Out Parade at Gibraltar Barracks.
The National Gliding Championships
The Award of the C.B.
The RAE Societies.
The Christmas Children's Parties.
The Christmas Tour
The RAE Messes.

11) 1984-1994   Royal Ordnance and Hunting Engineering.

The "Diamond" Committee.
The Royal Ordnance Factories.
    Hunting Engineering.
    Land weapons
    Air Weapons
    The Gulf War.
    Nuclear Weapons.
    The Atomic Weapons Establishment.
    Diversification and other Interests.
    Farewell to Hunting

12) 1970 Onwards   Other Activities

The Royal Aeronautical Society,
Brooklands Aviation,
GAPAN,
SBAC,
FEI

"When you have tasted flight, you will forever walk the earth with your eyes turned skywards, for there you have been and there you will always long to return."
Leonardo de Vinci.

Preface

I started writing these memoirs for the family. Within my lifetime, aerospace has developed from dreams and experiments of the pioneers to a major industry that has played a vital part in our defense and our quality of life. The story illustrates the diversity of tasks that were undertaken by scientists working in government, in industry and in the learned societies and their relationship with the technology and the political environment of the time. The question was how best to present this story. There are the separate scientific activities, the leading of group activities, the project activities and their objectives and the impact of all on family life. The mixes that I chose was to base the chapters on my career and, where appropriate within that chapter, complete the project story. I was often involved with some of them several times during my career. Where these tasks impacted on family life is also covered.

My father stimulated my fascination with flying both by his own experiences and the many visits to Cranwell. In 1938 he went to France on a visit with the British Legion. He came back somewhat depressed as his assessment of the French Army indicated that in the coming war they would be easily defeated. For me, going into the RAF was the natural next step even though I had seen many flying accidents in the locality particularly around Balderton Airfield involving Hampdens. It seemed vital to join the fight. Life would have been unbearable under a Fascist regime. After the war, the desire for peace dominated our thinking but there was a new threat from communism. Victor Kravchenco's book "I Choose Freedom" convinced me that we faced another evil that had to be resisted.

If we were to oppose these threats, we had to supply our forces with equipment that was as good as it could be within the financial constraints at the time. I, with many others, tried to contribute to this objective. In that process, I have made many friends. They have helped me along in every possible way. I thank them for all their kindness and support and, particularly my family, who in the early part of my career moved with me to new locations and in the latter period, had to put up with my numerous absences.

Most of the diagrams and photographs in the book are Crown Copyright. The exceptions are those in Chapters 1,some of 11 and all of 12. I am also most grateful to the Parliamentary and Scientific Committee, the Public Health Laboratories, the RAeS, GAPAN, SBAC, FEI, the Aldershot News, the Evening Standard, the West Byfleet Herald, DAWN, the Shuttleworth Trust, James Labochure, Alan Curtis and BAe Systems for permission to use some of their material and/or photographs. I should particularly like to thank my wife, Evelyn, Sir John Charnley, Wason Turner for their encouragement to complete this project.

I have worked for, with and managed some wonderful people during my career. My many friends have also helped me in many ways. I am sure that we had a significant influence on each other's lives. I do hope that they enjoy reading this book. Any views expressed in it are my own and do not necessarily represent those of MoD/HMG.

THK

Chapter 1

Where it All Began

The Magnus Grammar School.

In the late 1930's, my father, who was an Air Observer in WW1, used to take the family to the RAF Cranwell flying displays. To me, they were always incredibly exciting. Hawker Harts flew in formation with their wings tied together; there were aerobatics displays and bombing of targets on the airfield with bags of flour. With my pocket money I used to buy a magazine called Flying, The Popular Air Weekly for 3d. From the earliest days that I can remember, I have dreamed about flying.

The Head Master of the Magnus Grammar School at Newark, the Rev Campbell –Miller, started a school Air Training Corps Squadron, No.1260. It was very small, probably no more than 40 boys. Naturally I joined and we learned the rudiments of the theory of flight and air navigation. We went on summer camp to RAF Newton, where we were able to get some air experience flying in a Lancaster during a test flight. Despite our very small size, we were entered into those races in the Nottinghamshire ATC Inter-Squadron Sports Competition at the Nottingham Forest Football Ground that qualified for the Challege Cup. The racing all went well. But to win the Trophy, we had to win the last race. It was a 220, 440, 220, 440 yds relay in which I was to run the last 1/4-mile. At the end of the last 220yds, another team had a big lead. Their last runner was running in plimsolls and sounded incredibly flat-footed as he passed to start the last 440yds. I started some 30 yards behind the leader and at a pace that I knew I could not maintain. But I had to catch him up and

The Winning Team of the Nottinghamshire ATC Challenge Cup 1942

managed to do that after about 220 yds. I then ran at his shoulder for the next 150yds trying to get my breath back. I then put in a final spurt leaving him a few yards behind. I think that it was the most exciting and challenging race that I ever ran. Our trophy was displayed with the team Rev Campbell–Miller, Jack Oliver, Johnny Barker, Dickie Dwyer, Freddie Brooks, D.Dixon, myself and Ernie Pilsworth.

I thoroughly enjoyed life at school and at home. There was a challenge in almost everything we did. In the last year, I was a Prefect, Captain of Dean Hole House, Captain of Rugby, Vice-Captain of Athletics and the Senior Sergeant in the ATC. Being an all boys' school, our social life had to be organised. We asked permission from the Headmaster to organise a dance with the local High School. He helped by ringing the Headmistress and they agreed as long as we invited, as guests, a Master and his wife and a Mistress and her husband. We then called a meeting with the High School prefects. It was at that meeting that I first met a girl named Evelyn Hughes. We got on well together, despite the fact that she had lived her life in the town and I was a country boy. The school dances were a great success. (I was delighted to hear in 1998 at the 50th Anniversary Dinner of the formation of the Old Magnusians, that despite becoming comprehensive and co-educational, the dances still continue and are well supported.)

In 1942, we had visits at school from RAF Officers who gave talks on joining the RAF. One of the routes that they offered into the service was via a University Short Course of six months. During that period we were expected to do the first year of a science Course and the work of the Initial Training Wing of the Aircrew course in the RAF. After a stringent interview and medical at Birmingham, the Secretary of State for Air wrote me a welcoming letter. At this point Evelyn entered training to be a nurse and I became an aircrew cadet.

AIR MINISTRY,
WHITEHALL, S.W.1.
22nd June 1942

A MESSAGE OF WELCOME

from

THE SECRETARY OF STATE FOR AIR.

You are now an airman and it gives me great pleasure to welcome you into the Royal Air Force.

To have been selected for air crew training is a great distinction. The Royal Air Force demands a high standard of physical fitness and alertness from its flying crews. Relatively few attain that standard, and I congratulate you on passing the stringent tests.

Not only have you passed these stringent tests, but you have also been recommended for the commencement of your initial training without having to wait your turn for recall from deferred service. You are exceptionally fit now or you would not have been chosen. See that you keep fit. Work hard but live temperately, and make yourself proficient at your flying job.

In wishing you success in the Service of your choice I would like to add this: The honour of the Royal Air Force is in your hands. Our country's safety and the final overthrow of the powers of evil now arrayed against us depend upon you and your comrades. You will be given the best aircraft and armament that the factories of Britain and America can produce. Learn how to use them.

Good Luck !

Archibald Sinclair

SECRETARY OF STATE FOR AIR.

A Message of Welcome from the Secretary of State for Air

To Durham University

In October 1942, I travelled to Durham by train from Grantham to join the Durham University Air Squadron, Short Course No.4. At York, a young chap called Johnny Moss got on the train and sat in the same carriage. We soon discovered that we were both heading for Durham and to the University Air Squadron. We humped our cases from the Station at the top of one hill to the Castle at the top of another. Altogether 106 aircrew cadets reported that day. The courses were designed to carry through the initial training work for the RAF and keep the universities ticking over during the war. The cadets were divided evenly between University College and Hatfield and we all ate in the Great Hall in the Castle. Science students made up A and B Flights and Arts students made up C and D flights. Flights A and C flights were Castle students and B and D flights were Hatfield students. By a peculiarity of the alphabetic/numbers game, I found myself in Castle for living and all University activities, including the science course and in a Hatfield flight for RAF activities. My accommodation was on the floor second from the top of the Keep. The windows overlooked Palace Green where all our drill instruction and parades were held. Due to poor communication and the fact that I had my accommodation in Castle whilst all the Flight Instructions for B Flight were put on the Hatfield notice board, I often did not know of a drill requirement until I saw my Flight under instruction on Palace Green. On all occasions I stayed out of sight and nobody noticed.

We realised our good fortune in being registered as undergraduates of the university and, even more importantly to many of us, being launched on the path to a pair of wings. The beauties of the City of Durham itself were an additional bonus and a revelation to almost all of us. How we came to be assigned to Durham we will never know. Strict rules of behaviour for students were enforced. If you were unfortunately caught by one of the two university policeman, the punishment was either "gating" or a fine. One of these policemen two also doubled as the Drill Warrant Officer for parades on Palace Green. Students were expected to book out at the Porter's Lodge if they were out in the evenings or at weekends and book in again before 10 p.m. at which time the gates were closed and locked. I think that, exceptionally, on Saturdays the gate locking time was 11 p.m. Gowns had to be worn at all times when in town. It was an offence to be caught without one. If one was late returning at night, access to Castle had to be gained by climbing the lower ramparts and entering through the coal chute.

Our rooms were cleaned and our beds made on weekdays by a team of ladies. Our rooms had two doors; the outer one being left open most of the time. If the outer one was closed, called "oaked", the message to other students was that you did not want to be disturbed i.e. you were assumed to be working. Unfortunately, "oaking" did not debar the university police if they suspected that we had a *GIRL!* in the room. We all ate in the Great Hall. It was only about 100 yds from my room but the cadets in Hatfield had about 1/4 mile to walk. As expected in wartime, the food was very basic and my memory is that the rice puddings were rather like a white "blobby" jelly that was very sticky and quite inedible. Our minor protest was to pile the plates up together with the rice pudding on them and then squeeze them together until the rice exuded from the sides. The plates were then extremely difficult to separate.

The RAF intended that we should be well prepared for anything that might happen to us. Making us fit involved early morning parades (6 am) to march, often in the snow, to the swimming baths. There we were tested and instructed on swimming in freezing water. At other times, we went on cross-country runs and exercises in parts of Durham County with other service personnel.

The Science Labs were over the river, about 1 1/2 miles from Palace Green. I found the work there to be fascinating. I often spent the whole day there. We had exams at the end of course and I was delighted that I had done well. After the war, I used the results to seek re-entry to the Honours Physics School. Being the only fit young men in the area, apart from the "pongos" at the local army camp, we had a heavy responsibility of entertaining the ladies of St Hild's, St. Mary's, Neville's Cross and Whiteland's college in addition to the odd local lass. There was no shortage of females at the college dances.

On the last night we all admit that we danced the conga from the Three Tunns through the adjoining streets finishing on Palace Green but many of us emphatically deny having anything to do with the decorating the police kiosk and the equestrian statue of the Duke of Wellington in the square in the city center. When the course ended we all went on two weeks leave before joining the RAF proper! Although we did not ever meet again as a group, some of us met again at the Air Crew Reception Centre in London.

I reported to the ACRC- (Arcey Tarcey) in St Johns Wood in London in April 1943. The objective of the four weeks there seemed to be to fill both arms with injections of all types, which swelled up in the most painful way. The treatment for this condition was to have us out on runs through the parks, drill parades and scrubbing the floors with both arms to ensure that the injections were absorbed as quickly as possible. We were fed in the restaurant of the London Zoo. The march to the restaurant through the zoo and queuing in the dark of the morning and listening to the bird and animal calls was always an interesting experience.

It was with great relief when I was posted to Stratford-on-Avon for some further training in Navigation etc. It was there that I was introduced to "scrumpy" (raw cider) and my memory is that it was cheap but left you with a terrible headache. I was then posted to RAF Carlisle to do the flying grading course prior to the RAF deciding whether I was the right material to make a pilot or some other aircrew role. The aircraft were Tiger Moths. I fell in love with them straight away. I went solo in 7 hours50 minutes on the day before my 19th birthday. I thought it was wonderful having the aircraft on my own. Although I followed the instructions faithfully i.e. to do one circuit and landing, I could not resist the temptation to open the throttle again as soon as my wheels touched and do another circuit before landing and taxiing in.

From there, I was posted to Heaton Park in Manchester to wait for an overseas posting. The park, on which had been built some temporary huts and blister hangers, was in the centre of Manchester. We were billeted with local families. In September we were assembled in one of the blister hangers to be told where we were to be posted. As the names were called I gradually became more despondent as all my friends seemed to be posted to

Canada or the USA. Eventually my name was called. I was to be trained as a pilot and posted to Southern Rhodesia. Only one other member of the Durham course, Norman Dunning, was posted with me.

We boarded the ex-cattle boat, the SS Orduna in which the accommodation was rudimentary. In the open mess decks, we had our meals, lectures and slept. It was incredibly crowded. The options for sleeping were - hammocks hung from the ceiling, -lie on the tables or -lie on the floor. The floor was painted steel, the tables were somewhat narrow and the hammocks, not only had no sticks to separate them at the head and foot, but there would not have been enough space if they had. I tried a hammock. It was like sleeping in a capsule, which was so narrow that the arms were strapped to the sides and the legs were tied together. I tried the tables and fell off. I tried the floor and found that everybody walked over you during the night. From Liverpool, the ship headed north and then west out into the Atlantic where we joined in a convoy. Then we turned south and eventually entered the Mediterranean through the Gibraltar Straits. By then the weather was better and I decided that sleeping on the deck was the best option. A way of surviving the conditions of the voyage had been found. Because we were aircrew cadets, we were allocated duties, through the night, of guarding the sleeping quarters of a group of 40 Wrens.

On the way through the Mediterranean, we had a variety of alerts and, although some ships were attacked, ours was not. Some of the ships broke off to the south to go somewhere on the North African coast. Eventually we arrived at the Suez Canal and travelled south through the Bitter Lakes to Suez. There we stopped to take on supplies but were not allowed off the ship. We sailed on alone to Zanzibar. Again we were confined to ship. We sailed on to Durban. In the Indian Ocean the sky at night was brilliant and as the water passed by the ship the florescence in the sea was surprisingly intense. We had experienced rough seas before but this was the first time I had seen long ocean rollers. They were so deep and so long that the ship could be lifted amidships so that the bow and the stern, with the propellers spinning, were out of the water. The ship would then dive into the next trough and climb the next wave. Seeing other ships travelling alongside was just fascinating to watch.

The people of Durban were very kind to us. They appeared on the quayside with offers to take groups to see the port and to their home for a meal. After two days, the train arrived to take us on to Bulawayo in Southern Rhodesia. On the train we were allocated six to a compartment, stacked three per side. The bunks above, or the ceiling over the top bunk were within three inches of the nose. We ate, slept and had our meals in that compartment for the six days it took us to reach Bulawayo. The city itself was spacious. The streets were built on a rectangular layout and were wide enough to turn, without difficulty, an ox-cart pulled by eight oxen in pairs.

Relative to the previous five weeks, Hillside Camp on the outskirts of Bulawayo was spacious and airy. As a group we were not only pale but also looked like the troops that supported General Gordon in the Egyptian Campaigns fifty or more years earlier. What little money we had accrued on the journey was then spent visiting the native tailors to get new tropical kit made or, if the issue kit could be salvaged, modified to a respectably modern design. We acclimatised and, by late October, I was posted to 27 EFTS (Elementary Flying Training School) at RAF Induna, about 10 miles north of Bulawayo.

I first flew there on 25th Oct 1943 with F/O Williams as my instructor. On the first flight he was teaching me about the spin and the way to recover from it. Within five days and 5hours 30 minutes flying I had gone solo again. This time I took 40 minutes over my solo flight. It was a dream. The airfield was at 4500ft altitude above sea level and, although we were only 20 degrees south of the equator, the climate was very pleasant. The countryside was relatively scrubby and featureless, but the airfield was situated to the west of a railway line that ran north-north east to Salisbury and another which ran west north-west to Victoria Falls. Even if you could see no recognisable feature and were effectively lost, provided that you remembered which side of the railways you were, it was easy to fly in the direction of one of them until it was possible to see a railway again.

SUMMARY OF FLYING AND ASSESSMENTS.

TOTAL FLYING	HOURS			
	DAY		NIGHT	
	Dual	Solo	Dual	Solo
92.20 Hrs.	46.05	41.15	2.50	2.10

TOTAL INSTRUMENT FLYING 5.00 Hrs.   TOTAL LINK 11.00 Hrs.

ASSESSMENT OF ABILITY (Form 5011)
MARKS OBTAINED - FLYING TESTS % of 1000   80.7

ABILITY AS PILOT:

An above average pilot. Confident and thoroughly reliable.

ABILITY AS PILOT NAVIGATOR:

Above Average.

Date 17th. Dec., 1943.   W/CMDR.
Officer Commanding.
No. 27 E.F.T.S. INDUNA.

The Induna EFTS Flying Assessment

I enjoyed the course at Induna. The flying was excellent and I enjoyed the ground instruction both in lectures and in the link trainer where blind flying and blind approach and landing techniques were taught. I had one frustrating difficulty with the link trainer. When I put the link into a climb, my weight and the centre of gravity of it was too much for the hydraulic jacks controlling its attitude. The result was that when I started to climb with the nose up, the aircraft would flop back onto its stops and the speed would fall very quickly and it would stall. I learned to overcome this difficulty by leaning forwards as soon as I put the link into a climb. I passed out from Induna with very good assessments. The Wing Commander said that they wanted me to become a flying instructor immediately but, after some pleading, they eventually relented and allowed me to go on to 21 SFTS (Senior Flying Training School) at Kumalo where, I was told that I would have no option.

Kumalo was on the outskirts of Bulawayo due south of Induna. It was a very pleasant station. I started flying at Kumalo on 7th Jan 1944 and was awarded my "wings" on 26th

My Wings Photograph

May. Having completed another 182.80 hrs on multi-engine aircraft -total 274.60hrs. At the Passing Out Parade, the wings presented were about 2 inches across and we all thought of them as moth wings. Immediately after the parade, we all trooped off to the military shop to buy ourselves a brand new uniform displaying the largest pair of wings that we could find. We were photographed so that we could send the prints home. At the beginning of June, I reported to 33 Central Flying School at RAF Norton near Salisbury to do the Instructors Course on Multi-engine Aircraft. The course went well and I thoroughly enjoyed it. I passed out on the day after my 20th birthday. I left CFS Norton, maintaining, to my delight, my "above average" pilot assessment.

I was delighted to be back at Kumalo. I thought that it was one of the nicest and the best-located stations in Southern Rhodesia. The SFTS courses were 16 weeks, compared with the EFTS courses of 8 weeks. The flying instructors were split into two groups, one taking each intake alternately very 8 weeks. Each Instructor was allocated 4 pupils whom we instructed in pairs. One pair started flying at 6.00am whilst the other started flying at 9.00am. The latter started their ground instruction at 6.00am and the former at 9.00am. At the beginning of the course, we would be flying some 5 hours each morning. In the afternoons, any aircraft requiring testing was flown and the instruction with each pupil, completed in the morning, was written up. As the course progressed and the pupils went on solo exercises, the amount of flying that we did each day gradually reduced. Later in the course, night flying, cross-country training and bombing were taught involving much longer flights.

There were, on occasion, some stressful times particularly when we first sent our pupils solo. My most worrying occasion involved a pupil who today we would have described him as anorexic. He concerned me greatly. If the pupils had not gone solo after twelve hours flying, they were considered for testing and probable re-categorisation for some other aircrew role. We all tested each others pupils before sending them solo. I decided that a second opinion was needed and asked another instructor to give him a solo test. This was done and he sent him off solo before returning to the flight line. We both spent about 20 anxious minutes waiting for him to get safely onto the ground.

At this time, most of the pupils allocated to me for training were older that I was. At the end of each course, we had a passing out party when the hair was let down and the new pilots could tell us what they thought of the course and us as instructors. I was very surprised when one of my rather overconfident pupils accused me of lack of confidence in him. I asked him on what basis he had come to those conclusions. In the cockpit, we sat side by side, the instructor on the right. Everything we did could be seen by both of us. When flying, it was very hot in those cockpits under the clear canopy and I used to loosen my seat lap belt a little when we were not in the circuit. When we were on the final approach coming into land, I used to tighten my seat belt. If there was any loss of control by the pupil causing bouncing or in some way failing to land safely, it was our responsibility to take control. In those circumstances, it was essential to be strapped in tightly as the aircraft could be doing some very weird things. He had obviously noticed this movement of my hands. He was quite a good student, but I told him that anybody landing with him needed to be strapped in tightly. In the last four weeks of the course, we took the pupils, the aircraft and ground crew for two weeks of operations from a landing strip out in the bush. For that period we lived under canvas.

We employed many of the natives at Kumalo. They were mostly orderlies who looked after our rooms and clothes, washing etc. Many were cleaners of the aircraft. When we did a test flight in the afternoons, they often asked if they could come flying with us. Frequently I took one, sat him on the wing spar immediately behind me and told him to hang on. To give them a bit of a thrill, we banked into 70-degree turns when they feared that they were about to descend from the heavens. With the "g" during the turn they were, of course, quite safe and they never moved an inch from their seats.

There were other diversions to maintain our interest, i.e. one ageing Hurricane and a Cornell. With a suitable briefing and understanding of the systems, we were allowed to fly the Hurricane. I managed to fly it several times. It was a tremendous thrill. I must admit that aerobatics in it were a dream and with that amount of power in the engine it seemed that nothing was impossible. However, on at least 50% of my sorties something went amiss. On two occasions, the engine started to leak glycol from the cooling system into the exhaust. This generates a mass of thick white smoke and, without significant sideslip, it was extremely difficult to see out of the cockpit. It was a great relief to get it back on the ground in one piece. On another occasion the oil pressure dropped dramatically so I did a semi-dead stick landing.

In the European theatre, the defeat of Germany was well underway and the RAF began to think of restructuring after the war. V.E day arrived and we stopped flying for only one day as most of our output was destined for the Far East. However, it was soon announced that Kumalo was to be closed and I was posted to RAF Heaney, an airfield some 15 miles or so from Bulawayo.

I had not had any leave since arriving in Rhodesia. There were a number of invitations on the notice board from families to spend some leave with them. One was from a Welsh miner and his wife working at Wankie Colliery not far from Victoria Falls. It would be nice to spend a few days with them and then a few days at Victoria Falls Hotel. They gave me a very warm welcome and looked after me very well. They took me around to see the countryside and he took me on a visit to the mine. It was an extra-ordinary place. There was an 18ft seam of coal about 50ft below the surface. The entry to the pit was by a sloping railway and they mined it leaving about 30% of the coal in large rectangular sections to act as pit props. From there I went on to Victoria Falls Hotel, some 20miles away. The hotel was an oasis in wartime and we lived in luxury. The falls were even more impressive from the ground than from the air. The combination of the thunderous volume of water, the height of the spray clouds and the intensity of the rain within that spray were really breathtaking.

I transferred to Heany and the flying and instruction proceeded normally. Then I was selected to go on a Central Flying School Course at Norton to get my A2 Category flying instructors certificate. I started flying there on May 28th 1945 and finished the course on 27th June with a flying instructor grading A2 category. On return to Heany, I was transferred to the final testing of the pupils about to get their wings. It was not always a happy experience. By then the engines of the Oxford had been upgraded and variable pitch propellers fitted. It gave the aircraft much more zip but it seemed far from clear whether the old engine mounting structure was strong enough for the new task. At that time we began to loose aircraft because an engine became detached in flight. In that event, there was insufficient directional control to keep the aircraft straight. The result was that the aircraft crashed because of loss of control.   The rate of engine failure also increased and many more were landed in the "bundu".

The pupil's final tests often involved throwing the aircraft around somewhat. In these manoeuvres, one could hear the stresses in the structure, much of which was glued wood that had been manufactured a long time ago. Luckily, I only experienced the loss of the cockpit top panels and had two engine failures not too far from the airfield. My hairiest experience was at night, when immediately after lift off, the aircraft began to shake violently. The speed was right, the attitude right and the aircraft was climbing. I took over, and instinctively put down 10degrees of flap. The shaking stopped. I returned and landed immediately. In the event, it turned out to be a loose centre section of the wing. How that could have happened I cannot imagine. Because of these problems, the instruction in difficult flight conditions was not always as thorough as it should have been. The problems were usually revealed in the final tests. At that stage, all one could do was to give the pupil some further instruction and reprimand his instructor.

V.J. Day arrived and suddenly we stopped flying. Some of the pupils were within two weeks of finishing their course and getting their wings. They were not allowed to finish unless they signed on for three years. Out of 150 pupils only about six signed on. By that time I had flown some 1000hrs.

This left us with a station full of people with nothing to do. Everybody wanted to get home as quickly as possible, but it was clear that the shipping was not going to be available for some time. The activities had to be physical and mental- lots of sports and lectures in the airman's huts. Some of these lectures involved long discussions on the future of us all. It was here that I first met serious left wing politics. At times the discussions became so heated that they were difficult to control. I'm not sure that overall morale was improved by these activities. Many of us employed the airman to make, from scrap, some large wooden trunks to carry home some of the luxury goods, like tinned foods and clothes, that the folks at home had been deprived of for so long.

Luckily for me, after about 6 weeks I was posted from Heany to Cape Town on the first stage of the journey home. The journey of six days was much more comfortable than that from

Durban. We were four to a compartment, which, at least, gave space to breath. The Lakeside camp was at within sight of False Bay and about halfway between the Simonstown Naval Base and Cape Town. The advantage from our point of view was that a railway connected all three sites. Within a few weeks, a ship arrived to take us back to UK. On board, life was a luxury relative to the journey out. Four of us shared a four-bunk cabin and we ate in shifts in a mess room. Life onboard was often chaotic and from time to time, some of the passengers provided some entertainment. We arrived at Liverpool in November in the middle of a dock strike. To overcome this problem, we set about unloading all the baggage. We left the rest of the cargo for the crew to worry about. I was posted to Harrogate, where we had to spend some days before I could go on disembarkation leave.

After my leave and a few time wasting postings, I had to report to Ad Astral House. There they told me that, as I had not applied for a Short Term Commission, I was not going to be given a flying posting. I told them that I wanted to go back to university as soon as possible and would like to start at the beginning of the University year in 1946. They told me that the chance of my being demobilised before 1947 was negligible.

I travelled to Durham by motorcycle and saw Professor Wagstaff of the Physics Department. He agreed to apply to the RAF for me to be released for the autumn term. This meant getting a Class B release, which cost me 10 weeks paid leave. This contrasted with the treatment of those in the Fleet Air Arm, who were able to go to university on full pay. Luckily, I was able to choose my departure date to be just before term started. I reported to a big warehouse in Birmingham to get my demob kit. It consisted of a suit, a coat, 2 shirts, 2 vests, 2pairs of pants, 2 pairs of socks and a trilby or a flat hat. All the suits were hanging on extremely long racks and the process was to find the style you liked and then find one to fit. A few days later I was on my way to Durham.

Our Wedding- 28th Dec 1946.

In October 1949, I went back to Durham full of hope and determination. Little did I realise what a difficult first year it was going to be! I had my motorbike with me and therefore being accommodated at Lumley Castle at Chester-Le-Street, some 10 miles from Durham, did not present much of a problem from the travel point of view. I had a grant of £285/yr. The intake to the Honours Physics School was a mix of ex-service people and ex-schoolboys. It was evident in the first lectures and exercises that they knew what the lecturer was talking about and we did not. The problem reared its ugly head in Applied Maths where we were expected to solve the problems using differential calculus and whatever we had known a few years earlier had been lost in the sands of time and was not going to be readily retrieved. Our first request was for some extra tuition that would be sufficient to get us through the first term. It was a major test of concentration and willpower and several ex-service chaps dropped out at this first hurdle. Any tests that we had to undertake demonstrated the problem with absolute clarity.

The ex-schoolboys got 80% or so and the ex-service chaps got 20% or so. The Physics and Chemistry were much more manageable. In fact, it was the pleasure of doing Physics that provided the stimulation for pressing on with the course.

To add to the complications, Evelyn and I decided to get married Christmas 1946. Rationing of everything made the arrangements difficult, but both sets of parents came up trumps with food and drinks for the festivities. Getting anything that was special was difficult and, therefore, not so surprising that the bulk of the presents were butter dishes ashtrays and glasses. Neither of us smoked and could not afford much to drink!

I asked the Master of Castle for a Letter of Recommendation to use to help me find some accommodation in Durham. His letter said that this student is getting married and needs to find accommodation outside the College. I can only assume that he did not approve of married students! The grant went up to £325/year. Money was incredibly scarce. My clothing seemed to be corduroy trousers and a sports jacket. Every cuff and elbow was covered with leather strips. Before long I had rejoined the RAFVR, flying from RAF Acklington and Evelyn had a post as Staff Nurse in the Outpatients in the hospital at Dryburn. Both helped but still fell short of generating enough money for the full year. We had a very happy time in Durham.

For me, it was an opportunity that was too good to be true. I worked incredibly hard, spending as much time as I could in the laboratories or in the library during the day and with Evelyn in the evenings. In order to relieve the boredom for Evelyn, I built her a radio receiver in a cigar box. It had earphones enabling her to listen to the home service and light programmes, hopefully, without disturbing me. The scheme worked pretty well but when she burst into laughter, I felt deprived and had to ask her what the joke was? It never seemed to make me laugh.

There were several other married students at Durham. The wealthy ones had managed to buy or rented some miners cottages, whilst the rest of us rented rooms somewhere in or near the City. As one might expect the married students and their wives became friends. We had a few parties. The drinks were of the low cost variety usually drunk from used jam jars. The Master even entertained us, with our wives and children, to some tea parties in his garden. The babies of other wives eventually turned Evelyn "broody" and we agreed that we would attempt to join this happy throng - provided that Evelyn thought that she could cope with the extra work generated by a baby as by then I would be well into my final year. (Such plans were always a gamble. However, apart from a problem with incompatible blood types, Evelyn's, Rhesus Negative, and mine, Rhesus Positive, which meant that Evelyn was booked into hospital for the birth and the nurses delivered Patricia in hospital in the middle of a strike, it all went well.) It was very interesting that the student wives took full advantage of the pre and post-natal facilities built for, but not used by, the miner's wives. Evelyn's description of the sights of these enlarged and mostly naked ladies under the sun lamps drinking their orange juice and cod liver oil was always amusing and sometimes unbelievable.

In the holiday between the second and final year, I took a holiday job with PYE at Cambridge. It was also a very convenient location because I was able to use the Cambridge University Libraries to help me to write my thesis on "Radio Noise From the Sun." The work for the thesis, stimulated by Dr. Prowse, proved to be fascinating. The Thesis was a success and did a great deal for me in stimulating my interest in galactic phenomena.

In the final year, we received many presentations on job prospects from employers. I wanted to fly again or be associated with flying. The prospects in aviation seemed endless with the recent entry into the jet age and we had a very interesting presentation from a chap named Haugh who was Deputy to Snow of the "Two Cultures" fame. This spelled out the career prospects in government research in terms of its breadth and opportunities. I applied to the Civil Service for a Research post at RAE Farnborough. The first interview was in Durham. The second interview was in London. The third interview was at Farnborough. All agreed that I should go to Farnborough into Aerodynamics Department and to Aero Flight to do Flight Research. They offered me £380/year starting pay. This was even less than I was getting as a student. I wrote pleading that this offer was not enough and they replied offering me £430/ yr. allowing me two increments for 4 years military service. I decided to accept, as there were promises of getting a house to rent after 6 months with them. I then settled down to get on with the preparation for my final exams. For a while, I didn't see much of Evelyn and Tricia.

I don't remember too much about the exams except that we had two practical exams of 6 hours duration. They were split into two three-hour sessions between which we were able to eat our sandwiches and have a drink. In order to relax us somewhat, Dr Prowse would take us out for a walk. On both occasions, it was to the top of a hill and down again. I guess it was irresistible to sing a few verses of "The Grand Old Duke of York, who had ten thousand men. He marched them up to the top of the hill, and marched them down again."

It was during these exams that independent examiners interviewed us. They asked questions about the experiments we were doing and general questions on our overall knowledge. The second of these exams will always remain firmly in my memory. Every honours student was given a different question to answer. After the first exam, discussion with the others seemed to me to indicate that those likely to do best were given the most difficult problems.

My task, in the second examination was - "Given about 20 ft of plastic thonging, establish its main physical characteristics and design and complete some experiments that define them?" I first wrote down what its physical characteristics might be. It would be easy to design some experiments to measure them. Somehow, it did not fit my perception of the difficulty of the exams. I thought that I had a good chance of doing well in my overall results but this experiment was all too easy. I sat and looked at that thonging for some time without getting a glimmer of the trick in the question. I started to play with the last 3 feet of it to see if something peculiar happened with an increase in temperature. It didn't. I then started pulling it with a steady load and strangely it began to stretch much more than I had expected. When the load came off, it gradually shrunk again to its previous length. I tried sudden snatch loads. Again, to my surprise, it was extremely strong and resisted the sudden load almost rigidly. The more I played with it, the more interesting the stuff became. I began to feel much more relaxed about both the question and whether I had found the characteristics that would make it a difficult exam question. Although I did not have time to establish what it was chemically, it must have been manufactured from a compound containing long chain molecules that had a time constant of plasticity that allowed them to stretch under steady load but remain strong under sudden loads. I have always felt that this practical must have given me good marks in the totals for the final results.

To my absolute delight and that of Evelyn and my family, I was awarded a First Class Honours Degree in Physics. With that came a cash prize that enabled me to buy some books on aerodynamics and the theory of flight to help with the new job at Farnborough. I was not due to join Farnborough until the beginning of September and, as always, money was short. I trotted along to the Labour exchange at Durham to sign on for a little dole. They asked me what I wanted to do. I explained the position to them. They then asked if I wanted a job immediately. I said no. They then suggested that I put myself down as a temporary clerk as there was then no hope of getting a job in Durham. I gratefully complied. We stayed with my parents on the farm near Newark. I thought that it would be useful to keep the £3/week unemployment pay and so I went to the Newark Labour Exchange to transfer my unemployment payments to Newark. They said that I must see the Employment Officer before anything could be done. I duly waited and was eventually ushered into his office. Who should it be but an old school friend of mine, Johnny Barker. He asked me what I was doing wanting the dole? He explained that it was difficult because only there were only 6 people drawing unemployment pay in Newark. However, after some discussion, he agreed that, temporarily, there should be seven.

Chapter 2

RAE Farnborough-1949-1955

I motorcycled down to Farnborough on the afternoon of the first Sunday in September 1949 full of hope and expectation. By the time I reached Bagshot, the roar of aircraft overhead was very loud as the final flying displays of the Air Show were under way. I booked into the Grange Hostel and was allocated a room shared with Allan McCurrach, a Scotsman working in Radio Department. The room was small, with two two-foot six-inch beds in it with a cupboard and hanging rail for each of us. I reported into RAE Personnel Department the next morning.

After a few introductory talks and the usual signing of the Official Secrets Act, I was told to report to Radio Department. This instruction came as something of a surprise, but I was told that there were too many recruits for Aero Department that year and I looked like a good recruit for Radio Department. They allocated me to work in a group researching into Single Side-Band Techniques of Radio Communication. As always, I was prepared to give it a try, but I was unimpressed with the total recruiting system. After a few weeks in Radio Department, I was not thrilled with the work and started a series of interviews with my various bosses to get a move to Aero Department. I must say that they were very sympathetic to my position and took steps to arrange the transfer.

I arrived in Aero Department with a number of other recruits. The induction policy at the time was good, in that we were each allocated a series of two weeks experience in some of the main Divisions within the Department, with the last two weeks in the Division where you were likely to continue. I started in Low Speed Tunnels where Miss Bradfield reigned supreme. Alec Spencer was her deputy and under their guidance I learned to install models in the wind tunnel and measure their characteristics with accuracy and without mishaps (like the model breaking free.) Miss Bradfield also liked to ensure that those in her department had an ample social life. To this end, she arranged a few parties including all the new recruits and a number of senior members of the Department so that we got to know each other better. The parties very successfully achieved that objective.

Aero Flight

From there I went to Aero Flight, with Handel Davies Heading the Division. I was allocated to the Stability and Control Section under Joe Lyons. It was here that I settled down and started my first serious work. My net take-home pay was £30/month. The solution lay in doing more flying with the RAFVR located at Woodley airfield near Reading. There were times when I able to fly over the house in Cove and do an impromptu aerobatics display for Evelyn's benefit. The two weeks continuous flying every year was not only a real pleasure but also a considerable boost to the finances to bring our heads above water again. One particularly enjoyable attachment, in 1951, was flying as Second Pilot with Transair Ltd a small airline flying Ansons and operating out of Croydon. At least fifty percent of the flying was at night operating to Jersey, Guernsey, Brussels, Paris, etc. delivering freight, mostly newspapers.

Stability and Control Research.

I first worked with Keith Smith. His main task was studying the implications of powered elevator controls on the flying and piloting problems of the Lancaster aircraft and thereby the implications for the design of future aircraft using power control systems. The power system was an electrically driven hydraulic unit. It had an adjustable scissors link between the pilot's controls to the hydraulic input and the power driven connection to the elevator control system. With this system we could examine effects of varying the gearing between the pilot stick movement and the elevator movement and the pilots reaction to the variations of stick force with stick movement and the aircraft response to it. With this system, the stick force per "g", that is the force the pilot has to apply to the stick to generate acceleration in a turn or pull-up, can be varied in any way required. The stick forces can be arranged as one wishes making the aircraft extremely heavy to fly, or so light that the aircraft could be flown by one finger. Either extreme could be very dangerous particularly the arrangement with light stick forces as the aircraft could be so easily over stressed or wrecked by applying a "g" overload. In addition, the pilot needs to be able to feel when his stick is in a neutral position and that he can release it, to do other things in the cockpit without worrying. Arranging a "stick force" notch at this point to create the feeling that occurs naturally in manual control, whilst under power control, was another of the objectives. Despite the fact that there was only one set of controls on the aircraft, it was a great thrill for me to actually fly the Lancaster and to be able to experience the control environment that the pilot was testing and commenting on.

In parallel with this work, there was another Lancaster in the Flight on which research on servo-tabs was underway by Johnny Walker. This is a system where the pilot's stick is connected, not directly to the control surfaces, but to a small tab on the trailing edge of the control surfaces. The control surfaces are themselves free, (other than perhaps a weak spring system), and the controls are moved by the aerodynamic forces generated by the tabs at the trailing edge of the control surface. The Brabazon aircraft controls were based on this system and a fundamental understanding of all the parameters was needed to generate the research information that would provide the basis of a solution to any problems that might arise during the aircraft development.

Spinning Research.

For me, the great opportunity came with some reorganisation and personnel moves within the Department. Don Harper, who had been running the Spinning Tunnel, moved to do research in the Transonic Tunnel; Peter Bisgood, who had been doing some stability and control work and full scale spinning work, was to move to be the Chief Ground Instructor at the Empire Test Pilots School at Farnborough and I was moved to take over Spinning Research. This involved, for the first time, the responsibility for Model and Full Scale Spinning Research put together in Aero Flight. With Dick Dennis and Mary Steedman as my trusty team, we went to work.

The Spin.

Historically, the spin has been one of the most deadly conditions of flight. A spin is normally entered by first stalling the aircraft. One wing generally stalls first or more severely than the other. The aircraft then starts to rotate towards that wing and, if no corrective action is taken, continues to rotate in that direction in the stalled condition descending rapidly towards the earth. These steep spins with the spin axis well separated from the aircraft but with different wing tilts were relatively easy to recover from. The recovery from the flatter spins is usually much more difficult. In the early days of flying, many pilots died because of the lack of understanding of the spin, the important elements of the aircraft design that caused the aircraft to spin in different ways and the piloting technique for recovery. In conditions of aerial combat in WW1 and the policy of having no parachutes, many pilots were lost.

In the late 1920's and early 1930's, S.B.Gates did a great deal of research to understand the spin, the design of aircraft to facilitate recovery and the correct technique for recovery. It was probably at his instigation that the first spinning tunnel was built in the end of the BETA Airship Shed at Farnborough. The airflow was vertically upwards with the working section surrounded by netting at the tunnel sides and at the top and bottom to catch the model and minimise the damage to it. The driving fan was above the working section and sucking air through it. The tunnel was open-ended at the top and bottom, the air circulating around the rest of the hanger. In the 1950s, the rest of the hanger was used for stores and was probably kept well ventilated by the induced airflow. By the time I took over, the only changes,

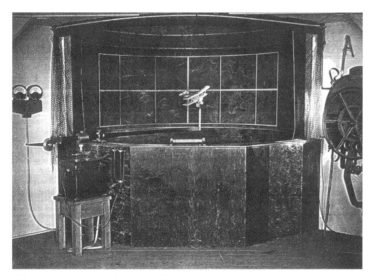

The Farnborough Spinning Tunnel.

from the early days, that had been made were to the rheostat control using a handle similar to that used for driving a tram and replacing the double wheel system. It was positioned on the right-hand side and a searchlight was placed on the left to provide enough light to take cine films of the model under test.

The technique, for operating the tunnel, was to hold the model in the spinning attitude in the tunnel whilst the rheostat was varied until the airflow was sufficient to carry the weight of the model. When those conditions were met, the model could be launched into the spinning mode and, with continuous adjustments to the speed of the tunnel, kept flying in the working section. The controls on the model were then moved, by a clockwork-delay timing mechanism, in the sequence required for recovery and that recovery observed and timed. If it were successful the model would dive into the netting somewhere, usually at the bottom, and be recovered, hopefully undamaged, for further tests. The models were about 1/20 scale and about 15 to 20 inches span.

A Model for Spinning Tunnel Tests.

They were built with great accuracy with movable control surfaces connected to a clockwork mechanism usually fitted into the side of the model.

Everything on the model was properly scaled for dynamic similarity to the real aircraft including the inertial ratios in pitch, roll and yaw. The required inertias were achieved by inserting small lead weights in the wings and fuselage as required. As the Reynolds number of the airflow at model scale was considerably lower then that in flight, there remained the aerodynamic scale effect that needed to be corrected. If it were not corrected, then the model would normally recover from the spin faster and more easily than the aircraft.

Gates solution attempted to correct for this discrepancy by applying a pro-spin force to the model by out rigging an aerofoil on a piece of piano wire positioned at a fixed distance from the centre line of the model. From the photographs of the model in the spin, the angle of this aerofoil could be estimated and the force on the model calculated. By varying the aerofoil size, the time required for recovery could be varied and by comparison with past experience, both model and full-scale, predictions of the spin and recovery characteristics made. This rather crude and empirical technique worked very well for many years.

In the new concept for the National Aeronautical Establishment at Bedford, a new large wind tunnel, pressurised to 4 atmospheres (45 lb./sq. in), was being designed and built that would overcome, by increasing the density of the flow, this aerodynamic scale-effect problem. The tunnel consisted of a vertical steel pressure -shell, some 80 ft high and 45 ft in diameter, with a fan, driven by a 1500 hp electric motor, and internal ducting to provide an airflow up the central working section and down the outer annular return circuit. The working section was 15 ft in diameter and could accommodate models with a span approximately 3ft.

When I reviewed the data available, it showed that there was a considerable volume of information on model testing from the wind tunnel but little available from flight where the spin had been maintained for long enough to reach stable conditions that could be satisfactorily compared with the tunnel results. AAEE Boscombe Down had generally tested in only four turn spins with little instrumentation in the aircraft. From this data, it was not even possible to be confident of the rate of turn in the spin. Eight turn spins were required to achieve any sort of satisfactory comparison, particularly if the ailerons were held in either the full pro-spin or anti-spin position. In the wind tunnel, the aircraft that exhibited the most violent oscillatory characteristics was the Lightning. As it turned, the nose rose some 20 degrees above the horizon and then returned to an angle some 70 degrees below the horizon. In addition, the rates of roll and yaw oscillated violently. From this violent spin, recovery is generally easy provided the pilot does not become disorientated.

In the tunnel, the model moved round the tunnel in a wide circle and behaved as if it was flying round the wavy edge of a saucer with steep sides. The aircraft rotated around its vertical axis some 4 or 5 times for each circuit of the tunnel. This flight trajectory is very inconvenient in the tunnel, as the model tends to fly into the netting around the sides before the tests have been completed. It was a bad omen for the future value of the new Bedford tunnel. Unfortunately, the design of fast transonic aircraft, with either stubby or highly sweptback wings and with inertia distributions dramatically different from those of past aircraft, was to change the whole approach to the problem and its solution.

(Several years later, Alan Clarke, my successor in charge of spinning had many problems with the Bedford tunnel. The fan was damaged during early calibration work and whilst it was being repaired, it caught fire and was again damaged. Eventually, the shell of the tunnel was used as a gigantic air bottle storing high pressure air to be used to generate a large diameter air-stream at 500 knots for 9 seconds to do research on pilot escape at high speed from aircraft and, in particular, the aircrew protective clothing and body restraints on the ejector seats.)

I decided that the first thing to do was to run a series of tests on current in-service aircraft and compare their spinning characteristics with previous spinning tunnel predictions. I seemed to have more power at that time, over the facilities made available for testing, than at any other time in my career. If I requested an aircraft for testing, within a month it would arrive. We would then install some instrumentation and start testing. We made the decision not to install anti-spin parachutes as it would take considerable time and we believed that the trials could be done without significant risk.

The aircraft we tested included: -
- the Percival Provost, which showed both the smooth and oscillatory spin depending upon the control positions. The oscillatory spin could be smoothed by the use of anti-spin aileron. Recovery from the inverted spin was rapid and satisfactory. It was evident that after a prolonged period of negative "g", of the order of -2 to -2.5 in the inverted spin, the positive "g" threshold during the recovery at which the pilot began to "red out" was much lower than normal; 3 "g" rather than 6 to 7 "g".
- the Jet Provost, which had the outer wings modified by sharpening the leading edges and giving some negative camber. This ensured that the aircraft went into a satisfactory spin suitable for training new pilots rather than a steep high "g" spiral of the original aircraft configuration.
- the Boulton Paul Balliol, which oscillated violently in the spin.
- the Vampire,
- the Meteor 7, etc.

During this period, I got to know many pilots very well. I was flying with them regularly and we were investigating many conditions that they had never met before. In addition, I was a guest lecturer at the Empire Test Pilots School. Our reward for this activity was to be invited, with our wives, to the end of course party. Many members of Aero Flight and members of other Departments in RAE were also guests. It was the party of the year. Very few of the chaps drove their cars home!!!

Escape from Spinning Aircraft

When we were testing the Provost and repeating, time after time, these rather sick-making long spins, it occurred to me that I had not seen any film of pilot escapes from aircraft in the spin or any study that explained the best procedures. I decided to do an experiment.

For many years the recommended technique for baling out of an aircraft in the spin, had been to exit the aircraft towards the inboard or trailing wing. This concept may have originated in the 1920s and early 1930s when the first series of spinning tests were completed. The aircraft of the period were biplanes, the usual layout being that the cockpit was behind or very close to the trailing edge of the wing. It was probably thought that the greatest danger was striking the tail during the escape from the aircraft. With the very low rates of descent of the aircraft in the spin at that time, baling out on the inboard side aft of the trailing edge of the wing probably ensured the safest escape. In these conditions there is little or no risk of passing anywhere near to the propeller disc.

To do the experiment, I had a number of balls of about 6 inches diameter of soft material, manufactured to represent the drag/weight ratio of a pilot. During the spin and sitting in the left-hand seat, I wound the canopy back sufficiently far to allow a ball to be put out into the air stream. On the first occasion, the ball was released over the trailing wing i.e. the port wing in a spin to the left as this was the recommended side for escape. To my absolute horror, the ball flew forwards straight through the propeller disc. We climbed up again and went into a spin to the right. This time I was close to the leading wing. I repeated the dropping procedure. This time the ball bounced along the upper surface of the wing and at about 3/4 span disappeared over the trailing edge. Escaping in this direction appeared to me to be the preferred way to leave the aircraft, if it was possible. These trials were repeated in the spinning tunnel using Provost, Balliol and Harvard model aircraft. A model pilot was fastened to the outside of the fuselage by the cockpit, and the model put into a sustained spin. The model pilot was then released from the leading wing and left over the trailing edge of that wing. The movements were photographed by cine camera. When the model pilot was released from the trailing wing, it passed very close to the propeller during an escape. These full scale and model scale tests were used to generate a new set of rules for escape from spinning aircraft.

Research into the effect of Inertia Ratio B/A on the Spin and Recovery.

When I took over spinning research, it was already evident that not only wing profile but also the ratio of the aircraft inertia in pitch "B" and the inertia in roll "A" made a considerable difference to the way the aircraft spun. The idea was conceived that on an aircraft like the Meteor Mk8, it would be possible to vary the inertia ratio B/A from 2.23 to1.12 by adding weight in the fuselage, 740 lb. in the nose and a similar load in the fuselage near the roundel, and then, after removing it, adding 250 lb. to each of the wing tips. There was an additional bonus using the Meteor as it had a Derwent engine set near to mid-span on each wing that could be used either to assist or delay recovery from the spin by applying thrust from one or other of the engines, as required. The loan of a Meteor Mk.8 was requested for a year of trials and before long it duly arrived.

As we knew that the aircraft had good recovery characteristics, we decided that the testing could be done without the installation of an anti-spin parachute. The engines had centrifugal compressors, which are very tolerant of the very poor intake conditions and could be used to delay or assist the recovery from the spin. The measurement of engine thrust at spinning incidence was extremely difficult, but essential if the pro-spin forces produced by the engines was to be used to increase the difficulty of recovery by use of the controls. The changes in recovery time with the pro-spin thrust of the outer engine was then used to assess the effects of the changes in inertia ratio and the resulting changes in control effectiveness during recovery. In order to get the most accurate measurements possible, the engine, with flight instrumentation installed, was tested in the cells at Pyestock. RR was consulted on the engine intake losses at high incidence and the in-flight estimates compared with model test results at RR.

F/lt Howard Murley (later W/Cdr Howard Murley DFC, AFC) was chosen as the test pilot for these trials. In this way, one pilot could accumulate the experience under the wide variety of spinning conditions to be expected. There is no doubt that his skill and enthusiasm played a great part in the success of the trials. In all, he did 91 flights and about 280 spins of eight turns duration. He started most of them at about 25,000 ft and then spun for eight turns before starting recovery action. By the time the recovery was completed he would be at about 15,000 ft loosing over 1000 ft in each turn.

When the fuselage inertia was high, the spins became oscillatory and very violent. The Lateral stick forces were extremely high and applied in violent snatches. In fact, they were so violent that they could not be countered even when the control column was gripped with both hands. When his elbows became very bruised, his wife Betty made him some elbow pads to soften the hammering of his elbows on the sides of the fuselage. On one occasion, the rivets sheared in the starboard aileron control run. This meant that one aileron was freely floating. The aircraft was still relatively easy to control at normal cruising speeds. However, when he reduced speed to the landing configuration, particularly with flaps down, the free aileron floated upwards. This aileron position had to be balanced by the other aileron being put in a similar position and he had to land the aircraft with the stick way over to the side of the cockpit with very little lateral control available to him. Despite these difficulties, our spirits remained high.

The results of these trials showed that when the moments of inertia in the fuselage i.e. the weight in the fuselage greatly exceeds that in the wings, then the rudder and pro-spin aileron are very important for the recovery. When the inertia distribution is mostly in the wings the elevators and anti-spin aileron are the most effective. These trials greatly improved the understanding of the characteristics of the spinning of modern aircraft and were described very fully in articles in Aircraft Engineering and Air Clues. The problem remained that to move the ailerons in the pro-spin direction to assist recovery from this out of control manoeuvre is against the natural instincts of the pilot. Even worse, it might be natural when in real difficulties for the pilot to apply anti-spin aileron thereby making a bad position much worse.

In the midst of these trials and at the outbreak of the Korean War, the MoD wrote to me inviting me to rejoin the RAF with my old rank to cover this emergency. I wrote back explaining what I was doing and suggesting that I was probably doing work of greater importance to the RAF if I stayed where I was. They replied a few weeks later with a slip of paper saying that my appeal against recall would be put before a tribunal. I handed the slip of paper to the RAE administration. Whatever they did must have been successful because I heard no more about the recall.

Spinning Research from Balloons.

In parallel with the Meteor trials, I recognised that the Javelin was soon to undergo spinning trials during its development programme at Glosters and then repeated during its clearance for service at AAEE Boscombe Down. We were concerned that the earlier wind tunnel tests of the Javelin would not have covered the effectiveness of the ailerons in the recovery from the spin or that the tests at such a small scale would adequately predict the characteristics of the spin and recovery. The earlier tests by Don Harper and Dick Dennis were repeated and extended using the normal recovery techniques i.e. opposite rudder and then stick forward plus the use of pro-spin aileron to assist the recovery. The reduction of the time required for recovery by using pro-spin aileron was dramatic. Any anti-spin aileron had a very adverse effect on the speed and probability of recovery. With the additional uncertainty of the aerodynamic scale effects of the delta shaped wing, we proposed that a new technique might be tried using a 2/11 scale model of the Javelin dropped from a balloon. The Reynolds Numbers of the large scale model would be much closer to those at full scale and thereby give us much greater confidence in the model results before the firms spinning trials were started. This programme was agreed and a model building company was contracted to build it.

The balloon trials unit operating out of the R100 Balloon Hanger at Cardington, was to supply the balloons and the crew to operate them. A new member was added to my team for these trials - Bill Verney, a retired RAF Squadron Leader Engineer, who was an expert model builder. He was made responsible for the design and testing of the control movement mechanisms and their timing controls and the installation of the recovery parachute. In addition, a cine camera team was required to photograph the spinning model during its descent, recovery from the spin and pullout and the opening of the parachute. Some of the members of the balloon operating team were interesting. A man named A.V. Bell had been one of the crew of the R101 when it crashed and burned in France. He and two other members of the crew were saved because a water tank, positioned above their station in the airship, burst on the impact with the ground and showered them with water whilst the airship burned around them. Holding wet handkerchiefs over their faces, they ran through the still burning skeleton of the airship and rolled in the long wet grass.

The model of the Javelin of over 10 ft span was mounted under the balloon carriage. Bill Verney and I ascended in the balloon carriage to launch the model into a spin and set the recovery control timers in motion. It is interesting to note

The Javelin Model Fitted onto the Balloon Carriage.

that we were wearing backpack parachutes. Ballooners normally wear parachutes with a line attached to the basket for opening the parachute. I was convinced that the most probable risk that would cause me to jump would be when the balloon was struck by lightning. With the balloon descending, I did not want to be dependent on this line to open the parachute. We normally dropped the model from a height of 3500ft to 5000ft altitude. To have the greatest chance of getting calm conditions suitable for the drops, it was essential to be airborne between 6.00am and 6.30am, which meant leaving Farnborough at about 4.00am in the morning.

With so much flying in spinning aircraft, we had been negotiating with RAE for sufficient flying pay to finance the extra life insurance premium needed to get some life cover should the worst happen. These negotiations were successful but they would not allow flying pay for ballooning because they argued that we were still connected to the ground. In fact, the cable caused the most dangerous part of the flight particularly in any gusting of the wind near the ground. In these conditions, whilst we were trying to get into the basket, the balloon swings like an inverted pendulum on its cable, from the ground to about 40 ft. and back again

The trials, which were very successful, were filmed for subsequent analysis. The conclusions of the report said- "The evidence, available from the 2/11 scale free flight model and a review of the tunnel results based on the latest scale effect data, suggests that the Javelin will have a slow oscillatory spin at low incidence with a rate of rotation of 0.9 radians /second. The recovery should be satisfactory by normal recovery action but if this fails putting the ailerons in the pro-spin direction (stick to the left in a spin to the left) will considerably improve the recovery standard. Great care should be taken to ensure that the ailerons are never held in the anti-spin direction otherwise the chances of recovery may be severely jeopardised. It is impossible to stress sufficiently how un-natural the *pro-spin aileron movement,* is to the pilot. He really needs to drill his mind to make this control movement which is against all his instincts."

At Gloster Aircraft, the spinning trials were to be done by Dickie Martin. He particularly wanted me to be available at the airfield when the trials were done. The many nights I stayed in the hotel room analysing the recordings of the days spinning were well spent. The trials were wholly successful.

In December 1955, after I had ceased to be responsible for spinning, S/Ldr David Dick was responsible for "A" Flight at Boscombe Down and undertaking the acceptance trials on the Javelin. The spinning trials were progressing well. One day at the end of a test flight he decided to do some high Mach No. stalls during which the aircraft flicked into a flat spin from which he could not recover. The accident enquiry, reported- "He had completed three normal spins when, before returning to base, he decided to explore the Javelins buffet boundary. At 40,000ft, the aircraft entered an unusual spin, and Dick realised that he was in serious trouble. Even so, as he lost control and the jet spun earthwards over the Isle of Wight, Dick continued to monitor the performance. Calmly and methodically he gave relevant information to a wire recorder. He did not eject until he reached 8,000 ft. —— By his coolness under extremely hazardous conditions, and by delaying his abandonment of this aircraft, much valuable information was obtained."

There is no doubt that the spin was very flat. At one stage, he selected the anti-spin tail parachute. It became trapped in the reverse airflow above the wing and, as a result, was quite ineffective. The aircraft crashed onto a fence. The incidence of the aircraft in the spin must have been very close to 90 degrees and the trajectory close to the vertical as a fence post went straight up through the wing and the aircraft remained fixed by it. As a result of this accident, the RAF reviewed their policy on the spinning requirements of all-weather fighters and decided that they should be designed so that they would not spin. The solution was to fit a stall warning system so that the pilot received warning, via a stick shaker, that the outer wing was in a stalled condition.

By this time, I was working in the High Speed Flight Test Group under John Charnley. Doug Andrews, who was working at the next desk, had considerable experience of the wing stall of the Avro 707 relating to the flow conditions over the wing under high altitude cruise conditions. His work had shown that without modification to the wings of the Vulcan, it would have been flying with the flow stalled in the area of the wing tips. Such conditions would cause a significant loss of performance as well as fatigue problems in service. As a result, the extended and drooped wing leading edge was added to the Vulcan design. He was given the job of designing and testing the installation of the stall-warning device in the Javelin. He chose a position on the upper surface of the wing roughly 50% span and wing chord. It was a small vertical plate attached to a micro-switch that detected the reverse flow of the stalled wing as it had progressed inwards from the wing tip with increasing incidence. When the stall had reached that point, the micro-switch closed and the stick started to shake to warn the pilot of the danger of a stall. At that point, the pilot felt no natural buffet or shaking of the aircraft to give him the normal warning of a stall.

My relationship with many pilots was extremely good. It is also an enormous pleasure when test pilots such as Bill Bedford and Hugh Merrywether express their thanks in their papers in Cockpit. Sept 1977 for my guidance and help during their flying.

The Comet Accident Investigation.

In the midst of the Javelin trials, I received an urgent request for trials to investigate some aspects of the accident to the Comets. In early January 1954, BOAC Comet "Yoke Peter" disintegrated close to the island of Elba. The pilot, radioing from 26,000ft, was broken off in mid-sentence. In April 54, another Comet G-ALYP was lost as it climbed through 30,000 ft. off Stromboli. All Comets were grounded and Sir Arnold Hall, then Director of RAE, offered the services of the Establishment in the investigation as there seemed something very obscure causing these crashes. Farnborough had experts on every aspect of aviation that could be applied to the task. He set up an accident investigation team in Farnborough. Political pressure was very high because the good name of British aviation was at stake. With the Comet fleet grounded, commercial pressures were also enormous. I can remember everyone in the establishment being asked to submit their theories on the possible causes of the accidents. From these suggestions, a large number of investigations were started in parallel.

Because the spinning activity made us the most expert group on dynamic scale modelling, the questions to us were: -

i) What would happen to the aircraft if the outer section of one wing broke off?

ii) What would happen to the aircraft if it broke up onto many pieces and what would be the trajectory and debris pattern on the ground?

For us, the big question was how do we do experiments that answer these questions? For both conditions, we had to be able to simulate the forward speed. To obtain that scaled speed, we designed a bungee catapult from which the models could be launched. To answer question i), the bungee launcher was set up on the top the R100 airship sheds and the model launched from there. As the model would have one half-wing missing, it would roll violently and therefore stay on a nearly straight ground track enabling it be caught on a large netting stretched out in the estimated impact area. Naturally we used quickly made silhouette models to test the system before the correct aerodynamic and inertially representative models were used. The trials showed that the rolling motion was much faster than initially calculated as, when the model rolled, the pitching moments of inertia caused the incidence to increase thereby increasing the rolling moment induced by the unbroken wing.

The Scale Comet Models complete and in a Break-up Form.

To answer the second question, models had to be built that represented each part of the aircraft in inertial, weight and drag characteristics. It took some time to make some assumptions on how the aircraft would probably break up. Then each part of the model was built separately and then assembled in such a way that on the pulling of one pin the model broke up into the various parts. We used the same catapult set up on a platform under a balloon. In this way, we were able to represent not only the dynamics of the break-up but also the speed and height thereby enabling us to get a representation of the debris pattern and its position on the ground.

We wrote Part 9 of the Report of the Comet Investigation. In the conclusions, it said -

"The wreckage patterns of the break-up models were reasonably consistent and have shown that: -

(a) If the break-up occurred at 30,000 ft, the main components were likely to have travelled a distance of 2 1/2 miles before striking the sea.

(b) These main components would probably fall within a circle of 3/4 miles radius.

Up to that time, the Navy had been unable to find the debris on the seabed. Off Elba, the seabed was sandy. Within two days of these results being available and knowing where to look, the larger elements of the missing wreckage were found. Once found, it could be dredged from the smooth seabed off Stromboli and lifted onto the ships. The aircraft, G-ALYP, was then reconstructed at Farnborough. Particular areas of interest are around and between the aerial windows on the top of the fuselage where the cabin failures were thought to have originated.

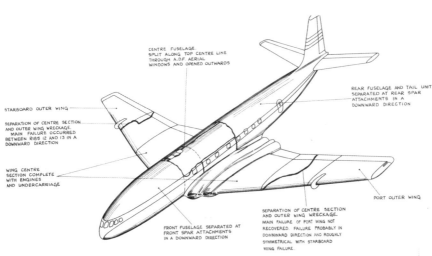

Location and Direction of the Main Failures— G-ALYP.

Another vital part of the investigation was the pressure fatigue tests in the water tank at Farnborough, which gave experimental confirmation of the likely cabin failure modes in the two accidents. In the early hours of 24th June 1954, it was discovered that the Comet cabin could no longer hold pressure and had burst. If that rupture had occurred in flight, the subsequent decompression would have ripped the aircraft apart.

The final report said: - *The examination of the wreckage shows that structural failure of the aircraft occurred in the following pattern. First, there was a violent disruption of the central part of the pressure cabin. Then the fuselage aft of the rear spar, the nose of the fuselage, and the outer portions of the wing fell away, all the action of downward forces. Then the main part of the wing, now a separate entity, caught fire. The fuselage, aft of the rear spar, complete with the tail unit, fell into the sea in a single piece, falling in an open-end first, tail aft attitude. The main part of the wing hit the sea in an inverted position.*

The flight plan that the aircraft was following would have brought it to an altitude of about 30,000-ft at the time of the accident. Supporting evidence showing that the break-up occurred at this height is as follows. Metallurgical examination of the burnt centre section shows that the fire was burning for about three minutes and calculations and model tests confirm that the time of descent of the centre section from 30,000 ft. altitude would be of this order. Tests on models confirm that a break-up of the kind outlined, at this altitude, would produce on the ground a pattern of wreckage similar to that found on the sea-bed near Elba, and that the motions of the larger parts would be of the type which would lead to impact with the sea in the directions which were thought to have occurred in the accident.

Considering the haste with which we undertook the Comet free flight modelling work, one cannot help but feel a touch of pride at the success of the programme and the correctness of some of the predictions we had made on how the aircraft might break-up and how best to simulate the likely ensuing sequence of events. The investigation led to a significant redesign of the fuselage structure by thickening the fuselage skin, changing the shape of the windows to near circular to avoid stress concentrations and introduced fatigue life testing as an integral part of all aircraft development programmes.

Some of the Fun

When I was instructing in Southern Rhodesia, my first Squadron Commander was a man named S/Ldr Porteous. One day he came to Farnborough to see me about the problem with the Auster. (He did not recognise me, but I recognised him.) He told me the story of a particular spin and the difficulties he had in getting the aircraft to recover. Somehow the story did not ring true, so I continued to question him for some time.

Eventually, he admitted that he had not given the full story. He was testing the aircraft at the full aft centre of gravity position. He had entered a spin and was unsuccessful in getting the aircraft to respond to his control movements to make it recover. He then started to move the controls back into the spinning position and then to the recovery position repeatedly to try to "rock" the aircraft out of the very stable flat spin into which it had settled. Unfortunately, his strong and heavy movements of the controls, particularly on the rudder, were such that the seat broke and he fell into the back of the aircraft. He scrambled back on his knees into the piloting position, fell forward over the stick onto the instrument panel at which point the centre of gravity must have moved sufficiently forwards for the aircraft to recover on its own. He pulled it out of the dive and flew it back to base on his knees. All in all, I would say a brilliant piece of piloting. The cure was a relatively easy one of putting some strakes about 6 ins wide and 2ft 6ins long in front of the tail plane at the top of the fuselage to increase the "damping" or side drag of the fuselage as it rotated in the spin with the result that the spin was at lower incidence i.e. steeper, and the recovery was much easier. Tiger moths had similar strakes fitted at about the same time.

The publication of the results of the Meteor trials and the Javelin trials resulted in my first invitation to present a paper at a Conference on Spinning at Dayton, Ohio in the USA. In the UK, there were three of us. In the USA, there were over 40 people operating three spinning tunnels and from the evidence of the papers at that conference, we were certainly as good as they were, if not ahead of them. I also managed to visit their facilities and talk in great detail to their teams.

It was in 1955 that another three-element move was arranged. Dennis Higton was to move to Boscombe Down, I was to move into his position in the High Speed Research Group under John Charnley and Alan Clarke from London was to replace me. For me, it was a great opportunity to enter another field of research that had considerable potential for some very exciting experiments in the challenging area of transonic flight.

Chapter 3

The Supersonic Flight Research Group

Transonic Flight Research-1955-1960

From the genius of Frank Whittle, the Jet Engine was developed presenting aviation with the opportunity of flying higher and faster than would ever have been possible with propeller driven aircraft. The Gloster E 28/39 was built followed by the development of the twin engined Meteors that were fast enough to catch the V1 flying bombs launched against London towards the end of WW2.

At the end of the war, expert teams were sent over to Germany to bring back both German scientists and German aircraft and equipment to further our own capability in the post-war period. These teams found that the Germans had also developed the jet engine but even more important had discovered that sweptback wings reduced the very adverse effects of shock waves on the flow over wings in high Mach number flight. They had also developed the rocket propelled Messerschmit Me 163 as a tailless design with swept wings and with a maximum speed of 620 mph. In December 1942, with the arrival of the Gloster E28/39 at RAE, jet engine experts from Power Jets were attached to a small group from Aerodynamic Flight Division to form Aero F/J. The group was lead by Ron Smelt and was tasked to explore rapidly, in flight, the further development of jet aircraft. To understand the many problems facing high-speed aircraft as the speed of sound was approached, flight tests were also conducted on a number of operational fighters and specially built research aircraft.

In 1948, Handel Davies was appointed Head of Aero Flight, and John Charnley, an early member of Aero F/J, was responsible for much of the Spitfire programme of transonic tests. The Spitfire achieved speeds very close to the speed of sound in steep dives and, on one occasion, the dive was so steep that the propeller constant speed unit became starved of oil, the propeller massively over sped and broke away from the nose of the aircraft, leaving it without power and the windscreen covered in oil. Squadron Leader Martindale made an astonishing "dead stick" and blind landing.

For exploring the various flight regimes, low speed tunnels and supersonic tunnels gave reasonably representative data. On the other hand, transonic wind tunnels were not very effective because the shock wave formed on the upper surface of the wing was reflected off the roof, floor and sides of the test section of the wind tunnel interfering with the airflow over the wing. Slotted tunnels helped, when they became available, but free flight models gave the best results although they were generally limited to very low incidences. The only way of overcoming the numerous problems arising at transonic Mach Numbers was to try it in flight. The RAF requirements were written in the form of an exhortation to achieve the highest altitudes, the greatest speeds and maximum "g" that aeronautical science could generate. This produced, for the RAF and Navy, many proposals for new fighters. As a bi-product of this policy, a number of swept wing prototypes for the next generation of fighters were produced that were surplus to requirements in the development programmes. In addition, a number of research aircraft were specifically designed to explore the performance and stability and control aspects of transonic and supersonic flight. The availability of these prototypes provided many interesting aircraft to test. I was delighted to join the High Speed Flight Research Group, headed by John Charnley.

I had been promoted to Senior Scientific Officer (SSO) in 1952 which at that time was the earliest that a scientist could be promoted (aged 28) to that grade. John Charnley was also an SSO. He could not be promoted to Principle Scientific Officer (PSO) because there was a bar on the promotion of scientists before the age of 31 and 1/2 years old. Such a bar did not apply to Administrators who could be promoted on somewhat better pay at the age of 28. It took 25 years for this anomaly to be eradicated and even then not completely.

Generally, we undertook three types of tasks -
i) Technical expert supporting HQ with the supervision of research aircraft being built, the BP 111A , BP 120, Supermarine Rocket Fighter, FD1, FD2, etc.
ii) Research on issues of transonic and supersonic flight.
P1052, P510, Hunter, Javelin, Wyvern, aircraft pitot-static calibrations, etc.
iii) Advice on project problems in industry.

Searching for "Fixes"

Shock Waves and Flow Separation Patterns ~ M= 0.95.

Many of the problems of transonic flight were generated by the shock wave formations, their movement over the aerofoil surfaces and the flow separations that ensued. Most of our fighter designs at that time were required to fly and fight in the high subsonic flight regime and therefore were affected by these shock wave and flow separations. Although swept wings delayed the onset of these problems to a higher Mach number, they did not prevent them. Further problems were created by the tendency of the air to flow span-wise along the wing surface creating the greatest boundary layer thickness and greatest flow separations near the wing-tips thereby causing pitch-up as the wing stalls.

Most aircraft of the early 1950's had wings with a sweepback of 35 degrees or so. As the incidence was increased and the stall approached, there was the usual buffet warning and then, with or without further warning, there would be a noticeable nose-up change of trim i.e. a self stalling tendency, needing a sharp reversal of stick force to prevent the nose rising further and the aircraft going deeper into the stall.

At low speeds, it was dangerous because this occurred, typically on approach and round-out for landing, accompanied with a significant drop in speed and a much higher rate of sink perhaps causing a heavy landing or worse. Only good flying technique overcame these problems.

In the high subsonic speed range, often involving high "g" manoeuvres, typically in combat, as the stall was approached, there would normally be a buffet onset. As the pilot pulled further into the buffet region, the nose of the aircraft would pitch-up into the stall. At best, the precise control required for aiming and firing the guns would be lost and, at worst, the aircraft would stall. In addition, on the Hunter, as the guns were fired the engine would flameout. The frustration of the pilot with the enemy in his sights must have been unbelievable. The cause of the engine flameout was the ingestion of unburned cordite powder from firing the guns into the intakes of the engine. Unburned cordite powder is very combustible and, when combined with the normal flow of engine fuel, so enriched the air/fuel ratio that the engine flamed out. That problem was eventually overcome by the combined efforts of Hawkers and Rolls Royce by cutting off the fuel to the engine when the guns were fired leaving only unburned cordite powder as fuel for the engine for that period.

Vortex Generators on the wings of the Hawker P510.

At the time, we had the P510 and the P1052 aircraft, both with swept wings, available to us. These aircraft were used to do experiments searching for a better understanding of and the mechanisms by which the flow separations on the outer sections of the wings could be reduced or delayed. Vortex generators, by producing a parallel stream of vortices near the wing surface, would re-energise the boundary layer by injecting of higher energy air into it and thereby delaying the flow separations that caused the pitch-up, were one of the lowest cost "fixes" available. The total pressure rake

P 1052 Rudder Fillet and Pilot Briefing

enabling the measurement of the depth of the boundary over that part of the wing is also shown. An alternative approach to the search for a solution to the problem of pitch-up was to produce, by adding a fence or significant wing discontinuity, both powerful vortex generators at about mid span.

The rear fuselage/ fin and tail-plane junctions also presented significant problems of shock-induced separations and buffet. Small experiments on the P1052 involving fitting fillets behind the tail-plane to reduce the shock wave strength in the fin/tail-plane junction involved extensive discussions with Ted Broadbent in Structures Department to ensure that the change in the mass balance of the control would not bring it close to the rudder flutter boundary. The briefing to the pilot often continued until he was thoroughly settled in the cockpit for a test flight.

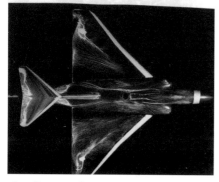

Javelin tunnel tests.

The Javelin suffered from both rudder buffet and reversal of rudder control at trans-sonic speeds. Wind tunnel tests using surface oil showed the position of the shock waves on both the fin and wings. A very successful modification was introduced and tested on the Javelin. It involved building up the top of the fuselage below the fin and rudder considerably reducing the strength of the shock wave and resultant separations to the extent that the problem disappeared.

At this time, the exchange of information with NACA of the USA was excellent. For me, absorbing the NACA reports, with those of Aero Department and the National Physical Laboratory, was an essential part of my job. NACA produced many reports of wind tunnel tests dealing with the study of shock movement on wings and controls in the transonic flow regime.

One day, Bill Bedford, then a senior test pilot with Hawkers and with whom I had done a great deal of flying, turned up in my office to discuss a problem they were having on the Hunter Mk 7. It was the twin-seat trainer Hunter on which it was particularly important that super-sonic flying could be demonstrated. It had a larger fin and rudder to counter the adverse effects of the larger cockpit on the directional stability. Unfortunately, this variant differed from all the others tested in that in a transonic dive the rudder " buzzed" i.e. it oscillated violently. They had tried physically locking the rudder control run so that it could not be moved. The rudder oscillations were so violent that the rudder control linkage broke. The company had involved the UK experts in transonic flow - A.Holder and H.H. Pearcy of NPL. The most favoured solution, at that moment, was to fit a robust damper at the root of the rudder hinge. The costs would be considerable and though this solution might have hidden the problem, it did not cure it. The residual stresses might also have seriously affected the fatigue life of the rear fuselage and fin structure.

I remembered some NACA tunnel tests and suggested that a solution might be found if the position of the shock waves on the rudder could be stabilised. In one of those NACA reports, the stabilising of shock wave position on control surfaces, when the control surfaces were moved, had been studied. Normally, the shock waves moved as the control surface was moved. The pressure on the control surface would then change and could cause both a rudder force and a rudder effectiveness reversal. At near sonic speeds, perhaps 0.95 to 0.98 Mach No., with the fin at zero incidence and the rudder neutral, the shock waves will be located on each side at part-chord on the rudder with separated flow behind each of the shocks. If the rudder is moved through a small angle to the left, with the view to yawing the aircraft to the left, the shock wave on the left moves rearwards towards the trailing edge of the rudder and the flow attaches on the left side. On the right side, the shock wave moves forwards towards the rudder hinge and the flow behind it detaches into a turbulent wake. On the left hand side with the attached flow, the pressure decreases; on the right hand side with the separated flow, the pressure increases; over these small angles the rudder effectiveness is reversed and there is a zero hinge moment position on either side of the rudder neutral position. If the rudder angle is increased further the effectiveness and forces reverse and become normal again. In these circumstances, the rudder has three equilibrium positions. The combination of the variations of rudder hinge moments with angle and the inertia forces in the rudder system could lead to a self-sustaining oscillation of the rudder, generating the conditions described by Bill Bedford.

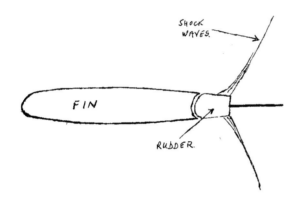

NACA Transonic Rudder Design

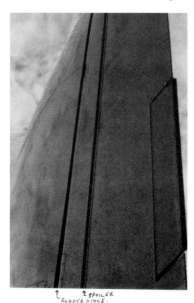

Hunter Spoiler Position

The Americans had tested, in the wind tunnel, a very peculiarly shaped rudder with a significant step in profile on both sides reducing the thickness of the rearmost 2/3s chord to a thin flat plate. The shock waves attach to the step in rudder profile and for small changes in rudder deflection, they stay located at the profile step. There are, therefore, no hinge moment reversals. The design that they had used was rather idealised and would be unsuitable for application to the Hunter. I suggested that we might try a thin spoiler put span-wise on the rudder at about 1/3 chord. We did some calculations on the probable depth of the boundary layer at that position and I suggested that the height of the spoiler should be about 1/3 inch deep. If this worked, the height should be reduced by 50% after each flight until it ceased to work. At that stage, the height should be increased to the next level up and fixed. Bill returned to the company with this suggestion. Everyone seemed to be against it. Pearcy said that it wouldn't work. Bill was determined to try it and had it fitted. The first flight produced some vibration but <u>no buzz.</u> The height was halved and the process repeated. The vibration decreased and there was no buzz. The spoiler height was reduced again by which time the vibration had disappeared and there was no buzz. The final spoiler height was 0.15 inch extending over the full span of the rudder. The cost of each aircraft installation was 17 shillings and 6 pence, probably the cheapest modification that was ever put on an aircraft. A similar modification was used on other aircraft including the ailerons of the Folland Gnat.

Supersonic Bangs

We had been testing various UK aircraft in the dive and reaching speeds that were just supersonic. Flying through to supersonic speeds on an aircraft with power assisted controls and a conventional fixed tailplane was difficult. The flight regime combined the rearward centre of pressure shift associated with the establishment of supersonic flow over the wings with the considerable reduction of effectiveness of the elevator and aileron controls combined with very high control stick forces. The F86E had fully duplicated power controls and the tail-plane was moved directly by the control column. The elevator at the trailing edge of the flying tail was geared directly to the tail plane and significant elevator deflections were used only in the landing mode. It was also fitted with full artificial feel for the pilot. We desperately needed experience on this type of aircraft. We asked USAF to loan us an F86E for a year. This wish was granted and one duly arrived. It was fitted with suitable instrumentation to record the most important parameters of flight condition and control position etc. At this time, supersonic speed could only be reached in a steep dive from about 40,000ft, pulling out at about 25,000 ft. The maximum speed reached was seldom greater than Mach 1.05.

The area of the trials was generally over Hampshire or Surrey. The area of ground affected by the sonic boom was generally small - a circle of about 1-mile diameter. It was not long before claims for damage began to be made. The reaction of people to the sonic boom and the damage caused was of considerable interest to Aero Department. "Chew" Warren led a small team that went out to see the damage and talk to those who complained. The noise varied from a mild rumble to a distinct double bang. One of the questions he always asked was "Did the glass (of the green house) crack or shatter?" The question always confused the claimants who often replied, "I'm a simple market gardener, how do you expect me to know whether the glass cracked or shattered?" Sometimes it was clear that damage had been done. Almost always the evidence had been cleared up by the time the claim had been made and the team had arrived. The RAE paid for the damage every time but gathered data that related overpressures with the probability of damage.

The Boulton Paul Deltas

Of the research aircraft, the two Boulton Paul Deltas were of special interest. Both aircraft were 45 degree swept wing deltas fitted with "elevons", combining the functions of elevators and ailerons, at the trailing edge of the wing. Both had conventional rudders. The BP111A was tailless, whereas the BP120

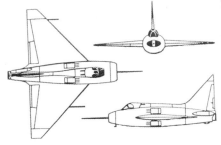
Boulton Paul 111A

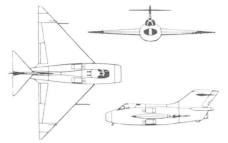

Boulton Paul 120

had a trimable tail plane placed at the top of the fin. The BP111A was first flown by Bob Smythe of Aero Flight on 10/10/1950 and a flight clearance programme undertaken by him with Jimmy Harrison and Jock Elliot of Aero Flight and Ben Gunn under the scientific guidance of John Charnley.

On August 6th 1952, Benn first flew the BP120 ominously known as the "Black Widow Maker". He was not particularly worried, as he had already flown the P111A, the handling of which he regarded as somewhat eccentric. After travelling three quarters of the length of the long Boscombe Down runway, he was dismayed to find that the jet had still not left the ground. His dilemma was- if he closed the throttle, a serious accident was inevitable; if he continued he may not get off the ground, which might be even worse. He decided to press on. Staggering and wallowing, the aircraft got airborne at the slowest rate of climb that he had ever experienced.

Three weeks later, on 29th August, Gunn flew the aircraft again. He was flying at 5000 ft over the south coast, when, after a burst of judder, there was a loud bang. The aircraft rolled crazily before diving towards the ground. He managed to recover control and get back to level flight but was horrified that the control column, rudder and tail-plane were all far out of position. At 3000 ft the aircraft went out of control and he ejected in an inverted flight attitude. The accident investigation afterwards concluded that intense flutter of the port elevon had caused it to fracture and break away from the aircraft. There was an immediate comparison with the BP111A. It was found that a strengthening plate in the centre of the structure between the elevon and its mass balances had not been included in the 120 (compared with the 111A) and this weaker structure had allowed the elevon to flutter in those flight conditions. It was also concluded that no modification was needed in the111A to avoid this problem.

The BP111A was painted yellow and became known as the "Yellow Peril". The plan was for the company to clear the aircraft through a very limited flight envelope but particularly up to about 500 knots indicated air speed. The aircraft was then to be handed over to me at RAE for research. John Charnley asked me "What my research plans were for the aircraft?" It seemed to me that there was a rather unique capability built into the BP111A, i.e. it had been designed so that the removable wingtips would allow the span and aspect ratio to be changed from a high aspect ratio wing with a pointed tip to a low aspect ratio wing with a significant wingtip chord. The aircraft was first flown with the intermediate aspect ratio wing tips fitted. We were also convinced that, at high Mach No., the flying characteristics were likely to be different in these three configurations and, in particular, the pitch damping at high aspect ratio might become small or even negative- a very dangerous flight condition. He agreed that this would be an interesting programme and that I should go ahead.

The first step was to extend the clearance envelope of the aircraft to altitudes up to 45,000 ft. so that the experiments could be carried out in the safest flight conditions in the region of 36,000 to 40,000 ft where the highest Mach Numbers could be reached at the lowest indicated air speeds. The first time 40,000 ft was attempted, the controls "froze". The pilot was unable to move the stick in any direction. He reduced the thrust by throttling the engine and the aircraft gradually lost altitude. After descending several thousand feet, the controls freed themselves and the aircraft became flyable in the normal way again. The power controls were operated by an electrically powered hydraulic system that was fitted across the top of the engine like a saddle. Consultation with all the experts on these systems did not produce a plausible explanation of the events at altitude. The only test chambers big enough to take the power control system, with sufficient space for the batteries required to provide the power for it and space for myself and others involved in the tests, were in the Institute of Aviation Medicine. They had an altitude chamber at ground level temperatures and a freezing chamber that could not be depressurised.

We chose to test in the cold chamber first. No matter what we did to the power control units, including soaking it at minus 56 degrees Fahrenheit for twenty-four hours or throwing water over it, the "freezing" of the controls in flight could not be reproduced. We next tried the altitude chamber. At 38,000 ft, the controls suddenly "froze" in a way that exactly replicated that experienced in flight. There happened to be a clear visual inspection port on the hydraulic fluid reservoir. We could see clearly that the fluid in it had foamed. If that condition were reproduced in the air at altitude, it would result in the control actuation jacks locking as the seals on the pistons would be so tight on the cylinder walls that the controls were, in effect, unmovable. "Eureka", we had found the cause. After consultation with the experts, the foaming was avoided and the required fluid flow at altitude restored by the enlargement of drainage hole in the accumulator cylinder.

We were now able to press on with the high Mach No. flights and, in particular, the pitch damping tests. The technique for the damping tests was to trim the aircraft hands off. Then, with the instrumentation running, give the stick a sharp knock forward. Under the spring forces of the artificial feel, it would return to the trimmed position. The aircraft would then oscillate in pitch and the motion and rate of damping would be recorded by the instrumentation. Normally, the pitch oscillation would be damped out in less than one cycle. I was concerned that, when we started to test the aircraft in the high aspect ratio configuration, the oscillations may be large, un-damped and perhaps divergent. The pilot would then attempt to stop them by the use of the control column. In this condition, there would be a real risk that his attempts would become anti-phase with the aircraft oscillations and the instability would be made much worse. There would be a risk of serious damage to both him and the aircraft.

The fastest and safest way to stop the oscillations was to slow the aircraft as quickly as possible. The

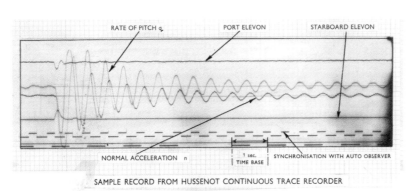

Hussenot Recordings of Pitching Oscillations.

throttle and air brakes were on separate levers. In a violent pitching environment, it is unlikely that the pilot would be able to transfer his hand from one lever to the other. We decided that the cockpit system had to be revised to combine the two functions into the throttle lever, i.e. the throttle as a normal lever and operating the air brakes using an electrical switch at the end of the throttle, operated by the thumb of the same hand. This modification was duly installed and the flight tests continued. A series of tests were made in which the speed was increased by 0.005 increments in Mach No. until we were close to the zero pitch damping expected. In that flight condition, the pitch damping had reduced to such a degree that it was foolish to continue further.

Prone Position Piloting

Increasing the 'g' level at which the pilot would suffer a blackout was a constant objective. One of the Human Factors research programmes of the Institute of Aviation Medicine, for prone-position experiments, included modifying the nose of a twin propeller driven aircraft, the Bobsleigh, built by Reid and Sigrest, called the Desford. There was another pilot position in the aircraft with a normal cockpit layout to accommodate a safety pilot. The prone pilot had a couch-like padded area to lie on extending from the top of the chest to the hips. There was also a chin-rest to relieve the strain on the neck caused by the weight of the head both normally and under higher "g" conditions. The elevator and aileron control was a normal "handle- bar" control with the normal elevator/ aileron functions situated just in front of and below the pilot's head and in a reasonable position for the arms. I guess the design was as good as it could be for flying in that position and avoiding the problem of blackout in high "g" conditions. I was excited to be able to fly it. The takeoff was normal and uneventful but, by the time we reached an altitude at which the aircraft could be put through some high "g" manoeuvres, I was already exhausted. My neck ached and my chest ached. Pulling high "g" manoeuvres was very painful on the neck and breathing very difficult. It was quite impossible to look around to ensure that the aircraft was not turning into the path of another. I found it impossible to pull nearly enough "g" to get the safety pilot anywhere near his blackout threshold. All the pilots flying it found that it was incapable of providing a solution to the blackout problem.

The other possibility for a change of cockpit configuration to increase the high "g" thresholds of the pilot was to install him in the cockpit in a much more reclined position on his back. B.Ae. Warton built a mock-up cockpit to give the pilots an opportunity to express their views on the advantages and disadvantages of this configuration. On a visit by the Chief of Air Staff, he was shown the mock-up and invited to sit in it and give his comments. He said *that he could not envisage any time that he would wish to go into battle, lying on his back, feet first with his legs apart.* From this point on, solutions to the blackout threshold problems were sought in other ways.

Promotion and The Air Efficiency Award.

During this period, John Charnley had been promoted to Principle Scientific Officer. I replaced him as Head of the Supersonic Flight Test Group as an S.S.O. In addition, after completing ten years of flying with the RAF, counted double for time service during the war, I qualified for the Air Efficiency Award.

The Fairey Delta 2

When I moved into the High Speed Research Group, one of my tasks was to provide the technical support for R.A.Shaw, the Superintendent of Research (Aircraft) in the Ministry of Supply. He was responsible for placing the contracts for research aircraft. One of those aircraft was the Fairey Delta 2. Every one to two months, I attended a meeting at Faireys, chaired by Shaw, when the progress and future programmes for the aircraft were discussed and agreed. The company team was headed by Bob Lickley, the Chief Engineer, and included Maurice Child the Chief Aerodynamicist. The meetings were always something of a tug of war between the company who wanted to get as much new technology as possible into the research programme and onto the aircraft and the Ministry team who wanted the aircraft into a flight test programme as soon as possible whilst staying within the limited funds available.

Two aircraft were in the building programme. The first, WG 774, was to be flown by Faireys and on which they would carry out the flight clearance of the aircraft and some of the research programme. The second, WG777, would be flown for the initial flights by Faireys and then passed to RAE for a flight research programme. Faireys felt that there was a window of opportunity to capture the world speed record to the credit of both the company and the country. The record stood at 822 m.p.h. and it appeared that over 1000m.p.h was possible with the Fairey Delta. The Ministry of Supply was not enthusiastic. Eventually it was persuaded to agree to the go-ahead provided that Faireys hired the aircraft and insured it for the flights.

The whole sequence of events is well recorded in Peter Twiss's book "Faster Than the Sun". It was a major piece of organisation by the whole Fairey team. They corralled Boscombe Down into supporting the aircraft, into the calibration of the static position error measurements on the aircraft and the analysis of the in-flight information recorded in the aircraft, the RAF into providing special radar units to control the aircraft during the speed record runs and the RAE Instrumentation Department into providing the means of measuring the speed of the aircraft with great accuracy. I was responsible for certifying both the measurements of the static error of the aircraft and that the aircraft had stayed within the specified height limits for the record run.

The overall task was enormous. All other record attempts had been made at heights below 1000ft where everything could be readily measured from the ground. For this record, a course, seen and measured from the ground by very accurate telescopes, had to be created at 38,000ft and the height during the runs maintained with great accuracy, measured by especially calibrated instrumentation in the aircraft. The skill of the pilot, the technology of the equipment on the ground, the co-operation between the radar controllers and the final corrections fed to the pilot by Gordon Slade from the timing cameras stretched the abilities of everyone and the technology of measurement to the limit.

The essential elements of the regulations and the way they were met were: -

1) The record attempt must be in a straight line over a course of 15-25 kilometres

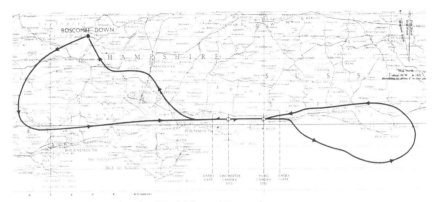

The Course Flown for the World Speed Record.

The course was 15,564.81 metres + or - 0.5 metres (below) measured by ordnance survey map, taking the radius of the earth as 6,371,227 metres. As the aircraft was about 7 miles high, the verticality of the cameras both cross track and along track was vital. The aircraft flying at a height of some seven miles would be photographed from the ground camera installation to fix the start and completion of each leg of the record attempt. All camera angles were measured by reference to the stars.

2) The altitude of the course is unlimited.

The competitor must maintain level flight within 100 metres tolerance at the entry to the course and over the course. The altitude at which the approaches to the course are entered in the two directions must also be within 100 metres. The height must be recorded on sealed barographs, and removed only by a steward.

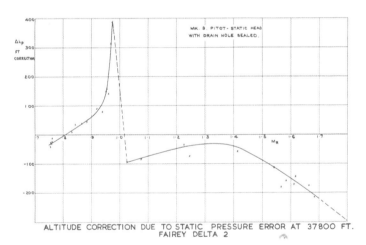

FD2 Static Pressure Error v Mach No.

The calibration of the position error as a function of Mach number was measured at 37,800 ft. If the aircraft was flown at constant height, the changes in height indicated to the pilot, as the Mach No. increased is shown. The requirement was that the pilot must fly the course in both directions and within 100-meters height (about 328 ft.). To ensure compliance was no mean task. Despite the changes in indicated height with speed, Peter Twiss achieved a slight but true climb throughout the runs.

3) The measurement of speed.
The aircraft shall fly over the course in each direction, and the speed adopted shall be the average to these two speeds. The two runs must be achieved within a maximum lapsed time of 30 minutes. No landing is permitted during the record attempt.

In truth, the aircraft was so limited on fuel that this track with the double acceleration could only marginally be achieved. The radar controllers put the aircraft on course as accurately as they could and alerted the pilot to the time to switch on reheat. At the entry camera station, Gordon Slade, by viewing the condensation trails, gave Peter Twiss final corrections to his course so that the camera crews had the maximum chance of taking the photographs needed for the timing. He also alerted him immediately the second photograph was taken so that the reheat could be switched off to conserve the very limited fuel available.

To attempt this record, the team needed practice. Everyone involved was doing a job that demanded 110% of his or her capability. The first runs were made on Thursday 8th March. There were difficulties in both height keeping and camera tracking. In addition, every time the course was flown there were a large number of people affected by the sonic booms. As the attempts continued, it was harder to maintain the secret that attempts on the world speed record were being made. Damage was being done to greenhouses etc., the press and public were curious, and objections were made to the repeated sonic bangs, because the public didn't know what was happening. Getting everything right at the same time began to appear impossible.

On the 10th March, the RAF Meteors had flown the course to check that all the cameras and timing equipment were in place and, on what could well have been the final chance, Peter Twiss took off feeling more tension than on any previous occasion. The first leg was flown and timed successfully, Peter flying about 9 1/2 miles in about 30 seconds. The view of the aircraft through the telescope was very limited and it was in the frame for a very short time. The second leg was flown successfully but the end-of-course camera had just missed the aircraft as it flew overhead. The aircraft could not have been far off the photographic plate and it appeared to be possible to fix the position of the aircraft at the time of exposure by the extrapolation of the edges of the condensation trails by comparison with other photographs taken as the aircraft was timed at the end of each leg during that same attempt.

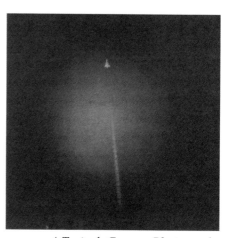

A Typical Camera Photograph.

Any such calculation had to be approved by the Royal Aero Club and the Federation Aeronautique International. I was able to confirm that the height requirements had been satisfied.

That morning, approval came through that the calculations of the aircraft position had been accepted and the record had been accepted at 1822 km/hour (1,132miles/ hour) on 10th March 1956. As Peter said "the sun travels round the world at 1000 miles/hour and he was flying faster than that"

Congratulations by Sir George Gardner, Director RAE.

The technique and equipment used and the detailed results obtained were written up on behalf of the Royal Aero Club by the RAE team, N.E.G.Hill, W Goldsmith and myself. We were eventually awarded the Diplome Paul Tissandier by F.A.I. Sir George Gardner, Director RAE, and Controller Aircraft, Ministry of Supply, Sir Claude Pelly, presented it to us. We could now all settle down to doing the research programme once again and the second Fairey Delta WG 777 soon arrived at Farnborough.

To Bedford

Supersonic Flight Research Group 1956

Aero Flight was under notice to transfer to Thurleigh at Bedford as soon as the facilities were ready. This responsibility for the timing of the transfer was delegated to each group. My main concern was that the new runway should be absolutely clean of any stones or debris. The FD2, with very narrow high-pressure tyres at 160 psi, would not take kindly to such objects on the runway and I had no intention of loosing this valuable research tool to such a problem. The Supersonic Flight Test Group included Doug Andrews, Bob Rose, myself and Peter Nicholas and on the

Aero Flight in about 1956.

back row, Cliff Spavins, Peter Haynes, Geraldine Holey and Freddie Dee. When we transferred to Bedford we had within the group, the FD2, a Hunter, and a Canberra. The picture above shows the whole of Aero Flight in about 1956. From the right, the sixth is Danny Lean, Head of the Low Speed Section, Bill Noble, the Flight Commander, George Zbrozek, Head of Stability and Control Section, Roger Duddy, Head of Aero Flight, Wing Commander Larson, Wing Commander Flying, and myself, Head of the Supersonic Flight Research Section.

In the mid-1950s, supersonic flight was still a very risky flight operation. It was not until the early 1960s that the Americans had the B 58 Hustler in production. Within two years in service, the Americans had lost 25% of the force. In the U.K., the data available on conditions in transonic and supersonic flight was very limited and, with particular regard to the FD 2, although the world record flights had provided some confidence that the aircraft appeared to have no unexpected vices in that flight regime, the extent of the coverage of the flight envelope was very limited and there were few wind tunnel tests within this speed range. In addition, the aircraft had no automatic stabilisation to help the pilot. As the control effectiveness changed with speed through the transonic speed range and with the change in indicated air speed, the pilot had to change, via a switch on the control column, the effective gearing between the control column and the ailerons and elevators. He was always busy and all exhibited considerable skill. With this valuable research vehicle the question was -What supersonic research could be done that would best further our knowledge of supersonic flight in all its aspects?

After some preliminary studies at Farnborough, the Supersonic Transport Committee was set up at RAE, chaired by Morien Morgan then Deputy Director (Aircraft) to explore the problems and the options in more detail. This Committee involved representatives of the aircraft and engine manufacturers, the airlines and government representatives. Our greatest contributions from flight research to these deliberations would be more information on the supersonic bang and the stability and handling of the aircraft under high lift/high Mach number flight conditions.

To obtain the maximum data from each flight, the aircraft was fitted with extensive instrumentation, much of it installed during the design and build programme. The port wing was extensively covered in tufts to show the direction of the airflow on the surface of the wing. These were photographed by a cine camera sited in the fairing at the top of the fin. The starboard wing had four chord-wise rows of static holes on the upper and lower surface of the wing. The static pressure at each pressure hole on the wing was connected through a rotating valve to a single highly accurate pressure transducer that was connected to the Hussenot data recorder. In any flight condition, the pressure readings obtained from the starboard wing can be compared with the photographs of the tufts showing surface flow on the port wing and the position of the leading edge vortex readily established. In addition, all other relevant flight conditions such as, speed, height, "g", control positions, control gearing, incidence, sideslip, etc. were recorded. All-in-all each flight produced a massive load of data for analysis and we were very dependent on the pilot marking the recordings at the times of most interest.

Since the Wright brothers first flight, it has been important to prevent flow separation over the wing and control surfaces. Flow separations have always meant some form of stall and danger to the flight of the aircraft. As wings were swept back, to ameliorate the effects of Mach number, flow separations tend to occur on the outer sections of the wing causing "pitch-up" and stalling. With highly swept wings, like the 60-degree deltas, the vortex flow separations begin at very low incidence and originate from the most forward part of the wing or wing fuselage junction. The research aircraft H.P.115, which had the facility to have the wing sweep changed in three steps from about 55 degrees, through 60 degrees to about 77 degrees, was designed to research the low speed flight regime. These low speed trials were particularly valuable in establishing that the separated vortices could cause a serious and divergent Dutch roll but that it could be controlled.

In the supersonic flight regime, these vortices existed within the complex shock patterns and their effect on the stability and control of the aircraft was unknown particularly in any dynamic situation. To explore this region of flight, taking into account the limitations on the aircraft particularly the

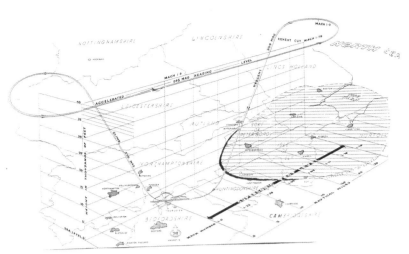

A Typical FD2 Flight Plan at Bedford

limited fuel, the flight-path/ Mach number profile used, kept the aircraft close to the airfield. When the test speed was reached, the aircraft was pulled into a turn in which the "g" was increased flight by flight to examine the flight characteristics with increasing incidence of the wings. The photographs of the tufts on the port wing taken in flight from the fin camera showed excellent correlation with the pressure measurements taken from the starboard wing and with the oil flow patterns studied in the wind tunnels.

The effects of sonic bangs were relatively unknown in 1955. There was some experience gained by the diving of transonic military aircraft up to just supersonic speeds. The areas on the ground affected were small, perhaps 1 to 2 miles diameter, but we already knew, from the flying at Farnborough, that the effects could damage buildings, disturb live stock and, at least, annoy people.

The area affected by the sonic bangs generated by the test flights from Bedford was much larger and there could be some focusing of the sound perhaps producing much larger pressure jumps in small areas. As can be seen, the total area covered could be 40 miles wide and have a length as long as the supersonic track of the aircraft. The focusing is likely to be in the region where the sonic bang first reaches the ground. This area was called the "cusp" and is shown as a thick black line. The pressure pulse could be greater than anywhere else along the rest of the track. Whether this happens depends upon the rate of acceleration of the aircraft, its height, the wind and wind shear with altitude and the temperature gradients in the atmosphere.

We involved the Building Research Station (part of the Department of Scientific and Industrial Research) in the measurement of over-pressures and assessment of the possible damage to buildings. With the kind permission of the residents, two properties, were fitted with vibration measuring instrumentation. My memory tells me that one of the buildings was the village post office and the building next door. A very large number of instruments were required to provide pressure recordings over a large area. In fact, only six were available. These were arranged at 1/4-mile intervals along the track ahead and behind the instrumented houses. A number of observers were stationed at various points in the neighbourhood.

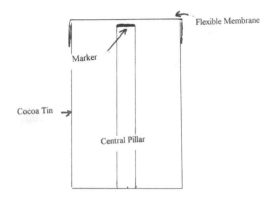

Cocoa Tin Peak Pressure Measuring Device

I think that it was Freddie Dee who had the bright idea to make some very cheap pressure recorders and distribute them over a much wider area. The essence of the design was to cover the open end of a cocoa tin with a thin rubber membrane. As it deflected under the pressure pulse, it would touch and be marked by the end of a central column on which a marking dye had been placed. The further the top of the central column was away from the membrane, the higher the pressure pulse required to make them touch. With a multiplicity of tins of different calibrations clustered together there would be a reasonable chance of getting an idea of the peak pressure pulse. The only readily available thin rubber membrane was a condom. Ordering a dozen gross of condoms through the civil service stores system was something of a challenge but they arrived with the empty cocoa tins and packs of peak pressure measuring kits were produced. Unfortunately, the experiment was not a success. We soon learned that, under a pressure pulse, the dynamic response of the membrane was different to its static response and it perished rapidly when exposed to sunlight.

Some special trials were flown at 10,000ft. The maximum shock overpressures measured at the ground were calculated to be between 2lb/ sq. ft. and 5lb/sq.ft., a value considered to be too low to cause direct damage to a building. (The atmospheric pressure at ground level is 2,110 lb/sq.ft.). However, the accelerometer records showed fairly severe vibration of the roofs of both instrumented houses and illustrates the tendency for there to be resonance between the sonic bang and some parts of the house structure. This feature might help to explain the apparent discrepancy between the actual and expected damage caused by the comparatively weak shock over-pressure. Minor damage such as broken windows and fallen plaster does occur occasionally although the incidence of damage is less widespread than commonly supposed. It was interesting that the public response to supersonic bangs varied more with the number per day than any other factor. One flight per day generated little, if any, public complaint; two flights per day generated a few letters and three flights per day produced a postbag of letters.

The complaints were varied and included letters from: -

a) people doing very delicate tasks, for example, surgeons doing operations where precision was required,

b) people working in old structures. For example, one complaint came from a school master who said that his school was about to fall down anyway, could we please do it after school was over for the day when there would be no risk to the children. A farmer complained that a wagon wheel had fallen off the wall of his barn and nearly decapitated him,

c) people in tragic situations including emotional letters, one of which said that his sick mother on being disturbed by a bang had rolled over and died.

d) farmers complaining of the disturbance to animals and chickens, which induced a tendency to romp and stampede sometimes causing premature birth, or in chickens a disturbed egg-laying pattern.

One day, whilst on a bus, Evelyn, my wife, over heard a group of women discussing the sonic bangs. One of them claimed that they had certainly changed the weather. This experience built up over a period of 5 years of supersonic flying involved not more than an average of 12 flights per month as experimental aircraft tend to have significant periods in the hanger for servicing and instrument calibration.

The supersonic trials using high "g" turns continued very successfully. The photographs of the tufts showed quite clearly the position of the leading edge vortex by the change in direction of the surface flow over the wing. The pressure measurements on the other wing gave considerable corroboration of this evidence. Especially set up wind tunnel tests gave some good flight tunnel comparisons.

The debate on the Supersonic Transport possibilities was very high in our thoughts. There seemed to be a real push for such a programme to try to gain a prime place in the civil air market. If it had government financial support, it would attract a great many advanced technology proposals, as everyone wants their technology to be a contender for the project. To this end, Werner Pinsker and I sat down together and put together a proposal to convert one of the FD2s to the most likely shape of a supersonic aircraft with "Ogee" wings. The main justification for the proposal given in the introduction to the paper was: -

" Before the final full-scale transport is built, it is most probable that two or three intermediate stages will be required. Gray's proposal for a glider is already being considered. This, when built, will increase the knowledge of the handling characteristics of such an aircraft through a fairly wide range of lift coefficient but will give little direct information about the take off and landing, and nothing about the characteristics in the transonic and supersonic speed ranges. Before the design of the transport is sufficiently advanced to start building even a scale model, a considerable amount of further theoretical work and tunnel testing has to be done. All this will take time. It would therefore be invaluable if another aircraft, already existing, could be modified to display the principal characteristics of the new platform to gain flight experience as quickly as possible to allow quantitative comparison with wind tunnel and theoretical studies. Such an aircraft need not be precisely the same as the final design and therefore work on the necessary modifications could be started very soon.

Over the past three months, possible modifications to the Fairey Delta, to enable it to fulfil this role, have been considered. Particular care has been taken to retain as much of the present aircraft as possible. By retaining the fuselage in its present form, the systems (fuel, hydraulic, pneumatic, and electric), pressure cabin and engine will be used in the new aircraft giving considerable relief in cost and design effort. "

The new items required were thought to be:
- two new wings, a new main and nose undercarriage and extension of the intakes to (we thought) above the wing (but in the design turned out to be below the wing). There would also be a requirement to extend the fuselage by four feet in order to get the Centre of Pressure of the wing and the Centre of Gravity in the appropriate places. Our preliminary design was followed by experiments on a mock-up of the forward structure to ensure that the view of the pilot would not be significantly changed by the outward extension of the nose to accommodate the front of the "ogee" wing.

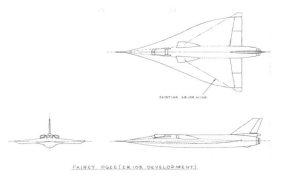

The Fairey Ogee ER103.

Mock-up of the Front Fuselage Modifications.

We published this paper on 28th August 1958. Fairys agreed that they would produce a short brochure on the structural aspects of the modification and give a preliminary estimate of the cost for submission to London. I felt that the proposal had to be sold hard in the Headquarters branches. To this end, I bought about twenty Airfix models of the FD2 and asked the workshops to modify one wing to show the comparison of the configuration of the two aircraft and distributed them to everyone that might influence the decision in London and Farnborough. It seemed to me that if they had it on their desks and looked at it every day, they would be convinced of the merits of the proposal. It was produced and (because of the changes of ownership of the aircraft industry) became the Bristol 221.

Flight at High Subsonic Speeds at Very Low Altitude

Military interest in the operation of aircraft at low altitude was growing because of the rapid development and deployment of surface-to-air missile systems and the vulnerability to them of aircraft operating at medium altitudes. Flying at high speed and at low altitude would reduce the probability of detection and engagement by SAMs and at the same time, relative to operating at higher altitudes, increase the accuracy of delivery of air to ground weapons. Unfortunately, there had been very little systematic work on the problems that might be encountered in this mode of flight.

A limited programme of work was set up using a Hunter Mk 6 and a Canberra fitted with a spring mounted seat in which a subject could be studied and give his impressions of the comfort and difficulty of carrying out his flight duties of controlling the aircraft and operating the levers and instruments in the cockpit. The Hunter was flown at about 200ft and Mach No. 0.9 over a number of level and hilly routes in the UK and over hilly terrain in Libya. The target flight profile was to be in level flight over the peaks and achieve as low an altitude as possible

Low Flying Routes Used in the UK

in the valleys. The overall objective was to minimise the average height above the ground thereby reducing the vulnerability of the aircraft by avoiding "ballooning" and greater exposure to SAM defences. The pilot was briefed on different flights to put a variety of levels of effort into achieving this objective so that the changes in the number and magnitude of the "g" excursions could be measured.

Due to fuel limitations, the average flight time at low level was limited to 12 minutes and the instrumentation was also very limited. Nevertheless, it was possible to obtain useful information on the turbulence level encountered, on the difficulties of contour flying over flat and hilly country, and on the many factors that affect the comfort of the pilot and his ability to perform his task efficiently.

The flight instrumentation consisted of a radio-altimeter, two fatigue meters fitted near the C. of G. counting excursions through 0.05, 0.45, 0.75, 1.25, 1.55, 1.95 and -2, -1, +3, +5, +7, +9 "g"

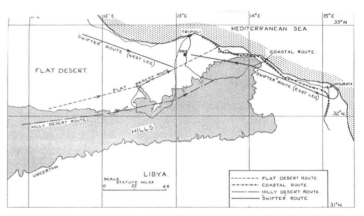

Low Flying Routes used in North Africa.

respectively plus the other usual flight parameters. Mike Goodfellow did most of the flying and, with Ted Nethaway, led the team to Libya.

In Libya, the pilot was asked to use a variety of efforts to fly at minimum height. To stay as low as possible required maximum effort. These results were compared with those obtained

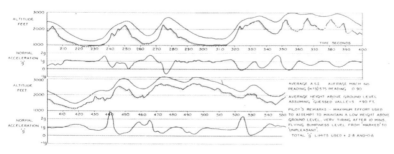

Flight Profile. Maximum effort

using a simulated terrain following system in the laboratory. In the laboratory tests, the average height above the ground could be reduced but at a cost of much higher average increments of "g" and about twice the frequency of "g" application. This higher frequency would have brought the frequency closer to the frequency of 4 seconds per cycle, known to be the frequency that produces the maximum occurrence of motion sickness.

Damage from a bird strike on the Hunter

The pilot's comments were an essential part of the investigation. Once the pilots were satisfied that the windscreen would protect them from a bird strike at high speed, they were able to undertake the task with some confidence.

The most important of the pilot's comments were-

a) the seat harness on the Hunter ejection seat was inadequate for this type of flying in that it was very difficult to get the straps sufficiently tight and even when this was achieved, they gave inadequate restraint to vertical body motion.

b) when flying visually, a clear view from the cockpit is extremely important. Apart from the structural members, the view occasionally became very restricted by the fouling of the front panel by insects and sea spray, and by pitting caused by sand and dust when flying over desert country. When flying in conditions of high humidity, condensation plumes formed on both sides of and above the cockpit. The plume above the cockpit seriously obstructed the view in turns.

c) the equipment should be designed to withstand a high level of turbulence. Difficulties were experienced during the flights due to fluctuation of the Air Speed Indicator and Mach Meter indications and the fuel transfer valves and indicators.

" Eye on Research.-Test Flight."

It was in March 1959 that the BBC first approached RAE with the proposal to do a programme on the supersonic flight-testing at Bedford. Their approach was straightforward. They wanted to concentrate on the pilot, Stan Hubbard, as he was such a character and a good expositor. A significant quote from their first proposal was "Female viewers will be primarily interested in the personality of Wing Commander Hubbard: as soon as he leaves the screen their interest will decline." Their other concern was to the length of the flight and the need to keep the total length of the programme to less than 30 minutes. Raymond Baxter, the excellent and well-known aviation broadcaster would lead the programme and we were delighted to work with him.

The problem with this type of programme is that everyone wants to be in it and I was determined to get as many as possible of the team, working directly on the aircraft, instrumentation and analysis of the flight data, into the programme as I could. The programme was unscripted. With Raymond Baxter, we worked out the sequence of events and the questions and answers flowed naturally. Some of the on-tarmac discussions involved Raymond Baxter, Ford Hutchinson the Air traffic Controller, Stan Hubbard and one of the aircraft fitters.

BBC Broadcast. Tarmac Discussions.

The sequence of events to be followed was-

- Flypast by the Lightning,

- An introduction by Raymond Baxter

- Meet Stan Hubbard and talk about the FD2.

- I briefed Stan for the flight and explain the instrumentation; this enabled me to include Cliff Spavins fitting the fin camera.

- Raymond Baxter then returned to Stan who was doing his final cockpit checks.

- The FD2 then gets airborne. The flight path followed was similar to that used for our normal test flights but the acceleration to supersonic speeds was earlier to ensure that the "bang" impacted on the airfield.

- Whilst the aircraft climbed, Philip Hufton, Deputy Director Aircraft RAE explained the supersonic bang phenomenon with the aid of moving diagrams.

- When the FD2 turned on to the supersonic track Ford Hutchinson did a countdown to the light-reheat point.

- I then predicted the time of arrival of the bang. On the first run I was right to within one second.

- Whilst Stan was doing the turn at high "g" he was reporting on the buffet levels. A film of the movement of the wing tufts as the leading edge vortex grew in size and the swing of the total pressure and temperature probe that swung through the jet pipe efflux so that the thrust of the engine could be measured.

In the Control Tower after the Flight.

- After that Philip Hufton explained some of the possible problems of lack of damping in pitch and yaw that might affect the stability and control of the aircraft at some Mach numbers and lift conditions.

-Wing Commander McCready explained what it was like to have command of the pilots doing experimental flight work of this type that covered the complete flight envelope from the lowest to the highest speeds.

- In a total elapsed time of about 28 minutes, Stan brought the FD2 in to land and the programme titles were run. We had a happy celebration in the control tower as the programme went so successfully.

Everything had gone well and the programme had been successfully received and taped in London. They then asked us to repeat the programme. When the programme was repeated and apart from my poorer prediction of the arrival of the bang, (I was 3 seconds early), it all went well. It was the second programme that was broadcast that evening on Tuesday 12 May 1959 from 9.15 - 9.45 p.m.

As the research programme continued, so did the supersonic bangs. In an effort to inform the public on the causes of the bangs and the importance of the research, Philip Hufton invited the Bedfordshire Times and Standard to a briefing- but the complaints still flowed in.

Career problems

It was at about this time that Roger Duddy, then Head of Aero Flight was promoted and posted to London. Frank O'Hara from the 3ft by 3 ft wind tunnel was transferred to be Head of Aero Flight. He was a very good chap but his appointment completely blocked my hopes of promotion to that job. I, therefore, contented myself with the thought that I had to move and hoped that such a move might be combined with promotion. Posts were advertised and I hoped to be boarded for them. The first to be boarded was for Head of Supersonic Free Flight Model Testing. It wasn't a job that I particularly wanted. My main competitor was Jim Hamilton. The Board was held in London, chaired by Walter Caywood with L.F.Nicholson and the Head of Civilian Management as Members. I think that it is important to set the scenario of the time to appreciate the likely thinking of those in senior posts at the time. In 1956, the Supersonic Transport Committee was set up at Farnborough based on work showed that "economic operation at supersonic speeds might be possible." The membership included seven aircraft firms, four engine companies, two airlines, and two government departments. Actually being engaged with flying within this speed-range meant that the progress of the work of this committee was of considerable interest to us. We were absorbed with many of the proposals that emerged from their work and, of the options being considered, liked the proposals for the slender deltas in the Mach 2 flight operational regime. Of course, there were many problems to which solutions had to be found. Even the airworthiness requirements had to be written and the operating costs were incredibly difficult to predict. The Committee warned that there was a danger that the operational problems would be neglected. To make the project viable the sales needed to be in the order of 300 to 500 aircraft.

The experience with the FD2 had shown that the public response to supersonic bangs was reasonable as long as the frequency was less than one a day. Three bangs a day even on only a small percentage of days generated a considerable number of objections. Although it would be flying higher, a supersonic transport would be very much bigger than the FD2 and the amplitude of the bangs would be much greater. They would have a larger spacing in the time of arrival between the shocks from the nose and from the tail of the aircraft, and it would depend on the natural frequencies on the buildings whether the resonance might cause more or less damage.

In addition, to have a hope of being an economic competitor to the high subsonic jets, the aircraft would have to do two return journeys across the Atlantic every day. It was very difficult to see how this could be done without one of the legs of the journey being at an incredibly inconvenient time of day.

Perhaps the number of complaints about sonic bangs that I had received over the previous weeks before the board had affected me but, on that particular day, I was not full of confidence about the future of supersonic transport. Quite early in the board the Chairman asked me "What sort of aircraft did I expected to be seeing in the sky by 1980?" I knew the desired answer to that question. It should have been " I would expect to see wave riders flying at Mach Five." This answer would have progressed into a series of questions about their expected benefits and problems to which I knew all of the answers. Instead the "imp" took over and I said, "I would expect to see 20 supersonic transports flying by then!" Well, the Chairman almost exploded. Luckily Nick came to my partial rescue by getting me to explain this extraordinary view of the future. I was not surprised that I did not get the job.

In circumstances like these, I always remember the story of the chap who was walking to work on a particularly frosty morning. As he walked along, he saw a canary lying in the road almost frozen. He felt that he must do something to help it- but what? Just then a horse came past and deposited a stemming heap in the road. Wanting to help, the chap picked up the canary and put it into the heap making sure that it was well wrapped in it with it's head sticking out of the top. As the canary warmed up, it started to sing. A dog came past. Seeing the canary stuck in this terrible heap, he decided to help. He picked the canary out with his teeth. Unfortunately, in the process he bit its head off.

Now the morals of this story are: -
It is not always your enemy that put you in it.
It is not always your friends that take you out of it.
It is sometimes better to be in it than out of it.
If you are in it, don't shout about it.

It is an interesting thought that my diabolical predictions turned out to be optimistic as only 16 production aircraft were ever built. Engine noise, supersonic bangs and costs proved to be the major problems.

I guess that I survived the board better than I thought because I was invited to interview for the jobs of running the Black Night Experimental team and to be Scientific Adviser to C-in-C Bomber Command and run the Operational Research Group. Harold Robinson was selected for the former and I for the latter.

I had to make my sad farewells to Aero Flight and we certainly had some wonderful parties to say goodbye. I was presented with the standard pewter mug. Aero Flight leavers were presented with one and they were treasured by all of us.

Before closing this chapter, I should like to mention the social life at Bedford. In all there were some 1100 people on two well-dispersed sites each division doing very different

The Inter-departmental Sports League.

types of job. Pulling together a social life was not easy. We organised and I chaired the organising committee of the interdepartmental sports league for several years. It included six sports with a winner's trophy for each sport. They were presented each year by Philip Hufton. That year we won the snooker trophy. Other sports were football, cricket, bridge, shooting and tennis.

CHAPTER 4

Operational Research- Scientific Adviser to Commander-in-Chief, Bomber Command. 1960-1964

For me, the appointment as Scientific Adviser to C-in-C Bomber Command was a great change, a totally new field and a challenge that I greatly welcomed. My knowledge of Operational Research was small and rested on its use in WW2.

Science had helped the military capability in warfare for many centuries. Archimedes made catapults for the Greeks of Syracuse to resist the Roman invaders; Leonardo de Vinci designed multi-barrel guns, tanks, submarines and parachutes. With the need to solve gunnery problems came the solution to the laws of motion and the foundation of dynamics. With the resources of the arsenals, researches in chemistry and physiology were made and the relationship between mechanical heat and work was discovered. With the need to treat the wounded came the creation of the medical hospital.

In WW2, science entered into the military field in a new role. The scientific method was applied much more widely, analysing the use of weapons and the relationship between the capabilities of those weapons and the tactics and strategy used in military operations. The first identifiable use of "Operational Research" was the study of the use of radar to provide warning of enemy air activity and to scramble our own fighters to intercept them. It played a vital role in the Battle of Britain and in the conduct of the air war generally. The story is well known.

One of the studies that always impressed me related to anti- submarine attacks from the air. At the beginning of the war, Coastal command had little success when attacking submarines. Changing the size and power of the depth charge had little effect. Even when it was dropped from a very low height, little success was achieved. Studies of actual operations revealed that the depth charges were set to explode at 50 to 150 ft. The settings were chosen on the basis that the submarine had a better chance of detecting the aircraft than the aircraft had of detecting the submarine. When the aircraft was detected, the submarine would dive and turn. The estimated depth that it would achieve by the time the aircraft could drop its bombs was about 100 ft. and that is why the setting was chosen. Studies, led by E.J.Williams and his team, revealed what actually happened. They found that, on one in every three or four occasions, the submarine had not detected the aircraft and was at or near the surface when the aircraft attacked. Thus the submarine could be attacked with much greater accuracy and the weapon effectiveness could be maximised by resetting the detonation depth of the depth charge to 35ft, the minimum depth to which it could be set. The number of successful attacks increased by 2 to 4 fold, a success rate increase that could be achieved almost instantly at no additional cost or training.

One of the keys to the success of an appointment of this type is their ready acceptability into the midst of an operational team. There is an immediate need to find mutual understanding of the objectives, responsibilities and risks. Perhaps my previous experience as an RAF pilot reassured them and was an advantage. I also had considerable support from my immediate deputies and all of the Operational Research Branch staff. Their main concerns were to ensure the Branch's continuous involvement in the most important issues in the Command and ensure that, in the continuous battle for status in an H.Q. with its long established hierarchical structure, that the Operational Research Branch continued to be held in high regard by the C-in-C and all his staff. For day-to-day activities, I reported to the Senior Air Staff Officer, AVM Paddy Menaul and I had direct access to the C-in-C whenever I wished.

The Operational Research Branch had four groups within it. The Training Analysis Group were led by an R.A.F. Flight Sgt and had 12 airmen of various ranks within it. Their tasks were to analyse the accuracy of the previous nights blind-bombing against targets assessed by the Radar Bomb Scoring sites and any additional trials such as Navigation Accuracy and Radar and Communication Jamming performance. The results had to be on the C-in-C's, S.A.S.O's, Group Captain Operations and my desks by 0900hrs the following morning. The Operational Planning Group operated using an Elliot 504 computer, on which we analysed the penetration capabilities of the aircraft against the Soviet Block, taking into account the latest information on the deployment of the Surface to Air Missile (SAM) systems, the deployment of the Soviet fighters and how both might be operated, with estimates of their likely success rates against the bombers. These results were analysed to give the probabilities of each aircraft reaching its target and successfully releasing its nuclear weapon.

Two very experienced Operational Research scientists led the other two sections. Ron Bruce dealt with counter measures, intelligence and airborne and ground radar capability. Hal Hood dealt with operational studies and weapons, guided and unguided. We were all part of the Department of the Scientific Adviser to the RAF and had a responsibility of liasing with other parts of the Scientific Advisers Organisation including H.Q. Fighter Command,

Training Command, the Bomber Command Development Unit and the Research Establishments. If the Branch was trusted by the RAF Officers in the Command, in the Groups and on the stations, the whole Command could be regarded as a very large laboratory in which any sensible experiment that might further the capability of the Command was possible. My relationship with both C-in-C's Sir Kenneth Cross and later, Sir John Grandy, was excellent. In addition to the normal day-to-day work, they involved me in the planning of the conferences and lecturing to the Command, Groups and Aircrews at the Bomber Command Conferences. I felt that it was a major endorsement of the work of the Branch and the esteem with which they regarded our work and, most of all, the frequency with which they sought our advice. The senior officers of the Command, photographed at the Commander-in-Chiefs Conference at North Luffenham on 14th and 15th November 1962, had a very close relationship. The front row, left to right are: - A.Cdre Dobell, AVM Menaul, AVM Dwyer, AVM Burnett, Air Marshal Sir Kenneth Cross, myself, AVM Rutter, Air Cdre Milligan, A. Cdre Johnson and Mr Campbell.

The role of the Command was the UK National Independent Deterrent and the 1960s were the height of the cold war. The deterrent posture at the time was based on the philosophy of increasing readiness as the enemy capability increased. At the time, it was largely an aircraft threat but medium range ballistic missiles were being gradually deployed. It meant demonstrating that the force could be alerted, be brought to a high state of readiness and had a target hitting capability that would deter the enemy. It was protected by Bloodhound SAM systems and was able to increase the difficulty of the enemy task and reduce his confidence in achieving it by being dispersed geographically or by being airborne. The task of the Operational Research Branch was to analyse both the enemy and our own capability and, by using scientific methods of analysis, recommend to the Commander-in-Chief actions that needed to be taken to achieve the above aim.

When I joined the Command, it had in its inventory of well over 100 V Bombers armed with various nuclear weapons and 60 THOR Medium Range Missiles with nuclear warheads. In terms of power to destroy, this capability was at its highest ever level and, with its ability to come to high states of readiness, was a powerful deterrent force. As the Mk2 Vulcans and Victors were introduced, the strength was due to decrease to some 84 front line bombers. The weapons were to be up-rated to Yellow Sun Mk2, a much more powerful variant, and Blue Steel, the standoff weapon, introduced.

The THOR Medium Range Ballistic Missile System

THOR was an Intermediate Range Ballistic Missile (IRBM) system with a range of 1,500 miles and a 1.44megaton warhead. The US government had agreed to provide them in 1957 to be in service until 1964. It was a dual key missile with separate keys held by a US officer and an RAF officer. The British key enabled the start of the launch sequence and the American key armed the warhead. Both would be required before the missile could be fired. However, the process of bringing the missile to readiness was a complex one. To bring the missile to readiness with the warhead stored in a central base would take 24 to 60 hours. In order to have the missile able to respond to a 24hr strategic warning, the warheads had to be fitted to the missile at all times. To finally prepare it for launch, the missile had to be erected and checked for serviceability and the fuel loaded. It could not be held in this state of readiness for long periods (more than ten days or so) because of the limitations on the life of the guidance system and the safety of the fuel. Some 65% of the force was kept at 20-min readiness.

A Combat Training Launch programme was set up in the U.S.A. The purpose was to demonstrate the capability of the launch crews, the adequacy of the maintenance and to build up the moral of the launch crews by giving them the opportunity to participate in an actual launch. The intention was to take one from an operational site, after a period of deployment in the UK, and, with the minimum amount of maintenance, launch it from Vandenburg Missile Test Centre. In general, missile systems were unpopular with the crews. Their unpopularity arose from the feeling that there was little end product that could be seen and appreciated, compared, for example, with aircraft that flew missions and returned.

As the missiles could not be launched from their operational sites, the trials had to be done from a trials site in California. Therefore, none of the ground support equipment could be tested in an actual launch. The ground equipment contained some 80 to 90% of the components of the whole system, and the test launch used different ground equipment, seriously limiting the value of the analysis of a test launch. Testing of the ground equipment in the UK depended upon repeated exercises bringing the system up to a variety of states of readiness. However, during the period between the receipt of the missile at Vandenburg and its launching, it was, for safety reasons, subjected to a much more intensive period of testing than would be expected on a squadron. After the missile was placed on its pad, it was subjected to a complete functional check and to a series of countdowns, which included at least one successful dry countdown, one single propellant countdown and one double propellant countdown. (All these trials, with the exception of the double propellant countdown, could and were done in the UK.) Various additional tests were usually required to check the telemetry, which proved to be rather troublesome during the programme. After the sequence of tests and countdowns, it was planned to hold the missile at standby for five to ten days. The effect of this intensive testing was difficult to assess, but it seemed reasonable to suppose that the testing affected the probability of a successful countdown more than the probability of a successful flight from lift-off to target. The first four firings indicated final countdown times of around 20 minutes. For the in-flight phase, the criteria of success required the missile to deliver the Re-entry Vehicle (R.V.) within a distance of the target consistent with a system error of 2 miles C.E.P. (Circular Error Probability) and the R.V. functions and the arming and fusing functions of the warhead had operated satisfactorily. All of these performance measures could be analysed successfully,

Despite the difficulties in the interpretation of the data and the differences between these launches and real operations, the launches were valuable in highlighting any deficiencies of the system and drawing attention to the action needed for modifications to rectify the faults.

Readiness

Over many years, the ability of Bomber Command to come to a high state of readiness had been developed, with the view to enabling the bulk of the aircraft force to escape a first strike surprise attack and be capable of a counter strike. When the threat was from manned bombers, early indications of aggressive intent of an enemy derived from intelligence information, combined with the deterioration of international relationships, was expected to give Command time to bring the aircraft to a high state of readiness. Early warning of attack, in terms of days, was expected from the intelligence organisation. Once brought to a high state of alert, warning from Fighter Command of the approach of enemy aircraft would give the bombers time to scramble. In addition, the bomber stations were protected by some 240 Bloodhound Surface to Air Missiles (SAMs). Bomber Command also developed a system of dispersal to other airfields such that no more than four bombers were on any one site.

On Oct 4th 1957, the Soviets put the first man-made satellite, Sputnick, into orbit. In 1959, a Soviet space vehicle orbited the moon and photographed the far side of it. These actions jolted the West into recognising the rapidity of the Russian technological advances and the potential capability that these launchers could be used for the missile systems with warheads mounted on their noses. It was not long before the Russians launched the SS-6 Intercontinental Ballistic Missile. In addition, on 1st May 1960 a US high altitude reconnaissance plane, the U2 was shot down over the USSR by an SA2 surface to air missile whilst it was flying at some 70,000 ft. Khrushchev was due to attend a summit conference in Paris at which Eisenhower was to propose an "open skies" policy so that at least some verification of deployments on both sides could be checked from the air. Any prospect of this proposal being accepted by Khrushchev was eliminated by the over-flight and shooting down of the U.2. He demanded an apology from President Eisenhower and the punishment of all involved. This was refused and the conference collapsed. These events affected the deterrent stance of Bomber Command. They required the progressive introduction of higher and higher states of readiness, standoff weapons and an airborne alert capability to maintain the credibility of the deterrent. New methods of intelligence gathering on the Soviet capability and the deployment of their systems were also required.

By 1960, the Rand Corporation in California, using new analytical techniques involving engineering, economics and military strategy, had completed a number of studies in the USA. The first study demonstrated that the air bases of Strategic Air Command could be vulnerable to surprise attack by ballistic missiles. The missiles could be used against political and economic centres as well as counter force targets to limit or destroy any counter attack capability. It is likely that if both sides had only a first strike capability, then this could lead to an extremely unstable situation, particularly in periods of high political tension. However, if the attacked nation possessed a force that could survive a first strike by having a sufficiently large force that was both dispersed or hardened, then it would have a second strike capability. If both sides had an adequate second-strike capability, there could be considerable stability, as there would be **no benefit** in making the first strike. Such arguments lead to the policy of Mutually Assured Destruction (MAD). Although there were many politicians and analysts who were not comfortable with it, they were unable to suggest a better alternative posture. This broad philosophy dominated operational thinking until at least the mid 70's.

When the main threat was manned bombers, in times of political tension, Bomber Command planned, depending on the intelligence information available at the time, to increase its state of readiness and then to disperse the force. Quick Reaction Alert exercises became part of the demonstration of the deterrent posture of the Command. The Stations had a requirement to meet the essential continuation training needs of the crews (some 35hrs flying per month), plus simulator training, exercises with their weapons and overseas exercises. The addition of Quick Reaction Alert Exercises required a considerable effort on the part of everyone on the stations and it was not surprising that the most effective ways of meeting these requirements was studied intensively by the Stations and the H.Q. staffs. At the Bomber Command Conference in November 1962, the C-in-C said there were Operational Readiness Platforms for some 100 aircraft available, generally in groups of four on 36 airfields. One of these airfields was R.A.E Bedford, my previous location, and I had seen them operating from there on numerous occasions. At 40 minutes readiness, the stance could be maintained for several days. The crews could get some sleep and any essential minor servicing on the aircraft could be undertaken. At 15-min readiness, the crews had to be fully ready in their flying kit. That state could be held for only a matter of 15 hours.

When the USA deployed BMEWS, the Fylingdales element of it was particularly relevant to Bomber Command as the warning times from IR Ballistic Missiles could become low as 4 minutes. At four-minute readiness, the crews had to be in the cockpit, all pre-flight checks done and able to take off within that time. On almost every exercise, the readiness states were increased to the four-minute level and the crews timed in their response to the order to scramble. Their record of achieving the 4-minute target was remarkable.

The C-in C commented that the effect of the introduction of the Quick Reaction Alert (QRA) that had led to integration with the plans of Strategic Air Command rather than co-ordination with them. SAC had maintained 50% of the force at readiness. It was accepted that Bomber Command could never achieve that level. However, the exercise experience had enabled our readiness states to be changed, as necessary, in a covert manner.

The Cuban Crisis

The Cuban crisis deserves a detailed discussion in its own right. Writing this script some 35 years later, it is possible to describe not only what occurred at the time but to augment this information with that available from the release of documents 30 years after the event.

The Cuban crisis lasted from the 14th to 28th October 1962. The first photographic evidence of the installation of SS-4 Sandal MRBM's with a range of 1250 miles and the SS-5 Skean missiles with a range of 2000 miles on Cuban soil was obtained by a U.2.flight on 14th Oct 62. The photographs showed canvas covered missile launchers and, although convincing to expert photo-interpreters, this interpretation was not self- evident to inexpert observers. Further evidence was obtained on subsequent flights and President Kennedy was informed on Oct.16th. Some members of the UK Intelligence community were informed on 19th Oct. President Kennedy informed the British Ambassador on 21st Oct. A rather obscure signal was received by the Foreign Office on Sat 20 Oct and a formal message from President Kennedy to the Prime Minister Harold Macmillan was received on Sunday 21 Oct. Bomber Command was already on a routine readiness exercise on 14th Oct. The Operational Research Branch was not monitoring that exercise.

These exercises were within the control of the C-in-C and his Operational staff and allowed some measures, which do not overtly display the readiness states of the Command, to be taken as a matter of routine. They did not allow its dispersal. These repeated and routine exercises in readiness states prepared the Command to meet the increasing alert states in times of crisis and thus enable the operation of Bomber Command to be efficient, automatic and well rehearsed.

The task at H.Q. was to obtain as much intelligence as possible so that, with instructions from the Prime Minister and M.O.D., appropriate action can be taken in terms of the Readiness State and the dispersal of the force. Much of the evidence given below from outside the Command is derived from the Kennedy Tapes taken in the Oval Office during the meetings the President held there. They were recorded, on President Kennedy's instructions, to provide an accurate record of the sequence of events, the actions decided upon and the views of those taking part. No one, perhaps other than Bobby Kennedy knew that they were being taped. His reasons for having such a taped record were to ensure that the record was exact. Its existence would prevent some of those taking part from "re-interpreting " their positions at some later date as had happened after the Bay of Pigs fiasco.

The record shows: -

14th Oct.- a U2 reconnaissance flight shows first identification of Soviet mobile missile sites in preparation in Cuba. The photographs showed 8 canvas covered missile launchers. These missiles posed threat to cities across America. The President + 12 people were involved. They considered all options, quick strike or negotiate etc. and the risks. There was a danger of Soviet action in Berlin. The military wanted to make a pre-emptive strike but could not guarantee taking out all of the missiles. The Soviets had given assurances that there were no missiles in Cuba only a few days before. More evidence was necessary.

18th Oct.- U2 flights showed new areas of MRBM sites. General Maxwell Taylor felt that the position could not be controlled by air attacks alone. Therefore, invasion was necessary if the military option was chosen.

19th Oct.- The President felt that the USA should not strike without warning. A way out had to be provided for Khrushchev.

The President leaned towards a blockade. The Committee split into the hawks and the doves. General Le May was probably the greatest hawk. The risk was that the Russians would take Berlin by force.

McNamara, Secretary of Defence, was unconvinced by the military arguments for action and argued strongly against his own military associates.

20th Oct.- The President developed a "diplomatic" cold so that he could leave the political campaigning for the mid-term elections and return to Washington. In reality, he wished to prepare a broadcast to the nation on the Cuban issue.

22nd Oct.- The President consulted the Congressional Leaders. The views of the hawks and doves rapidly polarised. The acrimony of that meeting was such that the President reminded those present that the person whose option was not chosen was in the best position. Whatever was done would be filled with hazard! That meeting was not a happy experience for the President as it took place immediately before his broadcast to the nation.

In his broadcast, the President announced the blockade. Cuba was to be quarantined of all offensive military equipment. Ships found to be carrying offensive weapons would be turned back. The President telephoned Harold Macmillan. Macmillan asked whether the President saw a way out? Had he spoken to Khrushchev? The greatest worry was that we had not acted with sufficient strength to start with. The President agreed to speak to Macmillan again.

In the United Nations, Adelaide Stevenson led a very successful debate against the Soviet Union. By mistake, the UK released the pictures of the missile sites before the US. The US covered up by saying that it intended to release them on the same day anyway.

Seven days had elapsed without the USA taking any significant action. President Kennedy was convinced that the blockade was the only sensible plan.

24th Oct- The American Forces Network (AFN) in Europe broadcast that the American forces were on full alert. Twenty-four Soviet vessels were on route to Havana escorted by submarines. Almost 480 US vessels supported the blockade operation. The destroyers were set to intercept the Soviet ships. The exclusion barrier was set at 500 miles from Cuba. Two ships were rapidly approaching that barrier.

Six ships, those of most interest, were seen to stop or reverse their course. The tankers continued.

Kennedy had further discussions with Macmillan. Macmillan congratulated him. He said that the turn back of the ships was a great triumph. It was suggested that they talk again on Friday 26th. Oct.

25th Oct.- US destroyer intercepts small tanker. The U.2.flights continued at low altitude to obtain better intelligence and were part of the demonstration to the free world that the US intended having these missiles removed.

There was further debate in the UN. It considered a freeze on all actions.

The USSR still did not admit that the ballistic missiles were deployed in Cuba.

26th Oct.- Macmillan suggested to Kennedy that he might wish to trade in THOR in exchange for the removal of the Soviet weapons if this would help in his negotiations with Khrushchev. At a Campaign for Nuclear Disarmament (CND) march in Bristol, the police were called into action to control the riot.

That evening Khrushchev offered, by letter, to withdraw missiles if the US declared that it would not invade Cuba. He then called for the removal of Jupiter from Turkey, Italy and Greece.

The President sent Bobby Kennedy to meet the Soviet Ambassador to tell him that if Khrushchev's proposal on the withdrawal of missiles in Europe became public then the USA could not agree to it. However, if it were not known publically, then the US could arrange for their removal. This was, in effect, a secret trade. (It was already planned that the THOR missiles in the UK were to be removed starting in April 1963.) Even though the US military did not know of this proposal, it would not have worried them, as they believe that neither THOR nor Jupiter was important. Jupiter did not work well and both were vulnerable to first strike attack and had a slow reaction time.)

In Bomber Command, some 59 out of 60 THORs were at 30 minutes readiness and the V bomber crews were at 30 minutes readiness at the main bases. Undoubtedly the crews on the stations felt that this situation was different and it was for real.

27th.Oct- Disastrous developments- The UN proposes a freeze on all actions and a U.2. is shot down over Cuba.

28th.Oct- 0900 a.m. On the radio, Khrushchev spells out measures that he was taking to preserve world peace. He ordered the dismantling and crating of the missiles in Cuba.

As part of the research for his book " Fail Deadly? Britain and the Command and Control of Nuclear Forces 1945-1964", in 1993 Dr Stephen Twigge wrote to me to ask for some comments on specific questions so that he could achieve a better understanding of the situation as seen in Bomber Command at that time. In that report, they describe Air Vice Marshal Menaul's account of the crisis as it unfolded. "He conveyed the impression of a taut but controlled situation. He said that the Defence Ministry in London and Bomber Command staff watched anxiously during this tense period and speculated on how America might react. By 20th October, the situation had deteriorated so seriously and rapidly that the Commander-in-Chief kept in continuous communication with the Air Ministry in London and with Strategic Air Command (SAC) H.Q. in Omaha, Nebraska." By contrast, Air Marshal Sir Kenneth Cross, the Commander in Chief of Bomber Command, said that "discussions with the Air Ministry were non-existent- and it was not for lack of trying. Sir Kenneth maintained that throughout the crisis, he was forced to act on his own initiative and, although he frequently tried to contact the Air Ministry, no response was forthcoming. The situation with regard to communications with SAC was much the same. Once the Cuban crisis started, there was no one at the end of the phone and there was no one there until the crisis was over. He suspected that this was deliberate."

My own view is that each officer was telling the story as he saw it. Their account of the events reflects their responsibilities and their personal approach to this very difficult situation. Air Vice Marshal Menaul would always want to be involved and I have no doubt that he was on the telephone continuously trying to get information from whatever sources were available. It would not have mattered to him whether his actions were immediately successful. He would have seen it as his duty to continue to try to gather information that would help to clarify the political and military situation. Sir Kenneth Cross, on the other hand, being responsible for the Command, would want hard information and instructions for his direct use. It was clear that no one was in a position to give him what he wanted.

As I was still working in the defence field in 1993, the Official Secrets Act conditioned any statement by me. I had to send my response through the MOD Security. I always felt that the US and UK deterrent postures were such that they would be perceived by the enemy as ready and effective and that, in these circumstances, if the enemy was logical and intelligent (and we believed that he was) and was knowledgeable of the risks in a nuclear weapons exchange, the Deterrent Force was unlikely to be called into play.

Lord Zuckerman recalls that within the Ministry of Defence, **no orders** were given to the C-in-C to change Bomber Commands alert state. That statement fits in with the readiness exercise that was in being and, under the C-in C's normal executive power, he was able to continue the readiness exercise and was able to implement measures that could be done routinely and covertly. Anything beyond that would have been overt and could have been construed as provocative and destabilising.

MOD records show that the first official instructions given by M.O.D. occurred at 11.00hrs on Sat 27th Oct. after the Chief of Air Staff, Sir Thomas Pike was called to Admiralty House for discussions with the Prime Minister. They considered what measures might be taken to alert the UK forces and Macmillan expressed his desire that overt actions should be avoided. Moreover, he did not wish Bomber Command to be formally alerted, although he wished the force to be ready to take appropriate steps should this action become necessary. After this meeting, Pike informed Cross that he should be on alert and that his key personnel should be available on station.

In 1999, I learned that from April 1961, one of the greatest spies working for the West was Oreg Penkovsky, run by Gervase Cowell of MI6. Penkovsky was a high level officer in the KGB and concerned that there was a great danger that Khrushchev's adventures would lead to war. [On an earlier occasion Khrushchev had said to Kennedy "if you want war, you can have war!"]. Penkovsky informed the West about the presence of missiles in Cuba and passed copies of the operation manuals of the missiles. Their presence was confirmed by the U.2. reconnaissance flights. He also revealed, the limitations of Soviet power, particularly their lack of atomic warheads and adequate guidance systems for their Inter-Continental Ballistic missiles (I.C.B.Ms). He also told the CIA that Khrushchev was bluffing and would not fire his missiles. [At the time there were only one or two people, including President Kennedy, who might have had this information.] Unfortunately for the West, Penkovsky was eventually exposed and arrested on October 22nd in the middle of the crisis. Under very fierce interrogation, he confessed the Cuban information immediately but did not reveal the full extent of his activities. Without a full confession with additional associated information, under the Soviet procedures at that time, he was shot.

With hindsight, it is worth trying to look at the possible gains and losses of this confrontation.

The world gained by –
i) the avoidance of war. I, personally, believed the deterrent worked and told Aberystwyth University so when they asked me.
ii) a direct HOT LINE telephone was provided for rapid communication between Kennedy and Khrushchev. (The means and mechanisms of communication throughout this crisis had been incredibly poor and slow.)

The Russians gained by—
  i) obtaining a promise from Kennedy that the US would not invade Cuba
  ii) obtaining a secret agreement to remove the I.R.B.M. threat, THOR from the UK and JUPITER from Turkey, Italy and Greece.

The Americans gained by-
i) getting the missiles removed from Cuba
ii) Experience in the handling of crisis situations. Kennedy was sure that he handled the crisis in the best possible way.

The Americans felt that although they had been successful, they should have got more out of the situation than they did. Over time, the Russians seemed to make significant gains. At the time of the crisis, they had some 42,000 troops stationed on Cuba. They used these experienced troops to train the Cuban army which was then used to destabilise and subvert the Central American and Central African states.

It seems to me that, in these situations, no more than one or two people know enough of the overall picture to be able to make rational decisions. One has to hope that they are "man" enough to take the decisions required at the time. It is an interactive mental combat to which very few can effectively contribute.

Bomber Operations

The use of air power is both complex and controversial. The strategy and tactics follow the general rules of war and involved
  i) air to air battles
  ii) support for the operations of the Army and Navy,
  iii) attacks against infrastructure targets, which can, not only cause damage by them, but also can, reduce the enemy's will to continue the fight.

It was during the Boer war that balloons were first used for intelligence gathering on troop movements and for artillery spotting. Aircraft continued this role in WW 1, initiating the first air-to-air battles. They also made the first experimental steps in strafing and bombing. The first air to ground radios were fitted to report the intelligence observations. The Spanish War saw the beginning of the development of "Blitzkrieg" air tactics in support of the army. It demonstrated the flexibility and speed with which air power could be deployed and the concentration of firepower that it could bring to bear.

World War 11 saw a rapid development of the use of air power enabled by a rapid exploitation of technological developments. At the beginning of the war, navigation accuracy was poor and finding targets was very difficult. Some estimate that perhaps only 10% of the bombs dropped in 1940 and 1941 were on the targets. The deployment of ground beacons generating a network of signals that were interpreted in the aircraft to give it accurate position information (The system was called Gee). It provided an essential element in the success of the 1000 bomber raids in1942. With the development of a beam system called

Oboe, the pilot did not even have to know where he was, as warning signals would tell him that he was near the target and when to release his bombs. As the aircraft could bomb from within cloud, there was much less risk from the fighter or anti-aircraft defences, losses rapidly reduced in some cases to as little as 1/4 % per sortie. However, these beams could be bent and the aircraft guided to an open country target as was demonstrated by our own bending of the German radio beams.

With the invention of the magnetron, giving the possibility of building radar equipment with wavelengths of 10 cms or less, airborne radar was developed giving a picture of the ground to the navigator in the aircraft. Urban areas stand out prominently and line features such as rivers and coastlines can be clearly identified. The equipment was given the code name H2S (which everyone knows is the chemical formula for hydrogen sulphide and smells like rotten eggs). It enabled bomb aiming to be done direct from the aircraft at night and in or above cloud thus giving much more flexibility to the operation of the aircraft. It also provided against any target, an accurate measure of height above the ground.

Numerous countermeasures, such as Window (chaff), clouds of silver paper strips that reflected their radar beams and could be mistaken for aircraft, were also devised to deceive the enemy radars, warning devices to alert the bomber crews of eminent attacks by fighter aircraft were also produced.

There were many developments in bomber capability between 1945 and 1960 when I joined Bomber Command. Our task was to study each element separately reporting on them to the H.Q. staff and then bring them all together in an assessment of the strengths and weakness of the whole force The results were integrate into the likely success rates of a bombing attack should the deterrent posture fail and the force be called upon to fulfil its retaliatory task. The success of the deterrent posture depends on the enemy perception of the risk that he would run should he take pre-emptive action against the UK or its Allies and whether a pre-emptive attack in any form was or was not worthwhile. I will deal, in turn, with each of the elements that would affect the outcome of any confrontation.

Intelligence

At the end of WW11, very little was known about the Russian Industrial and Military infrastructure. When the IRON CURTAIN was erected, it became almost impossible to gather any information. Travel was very restricted. What was known in 1950 was extremely limited. If ever it became necessary to attack Russia, information on the structure of targets and the location of defences particularly, surface to air missile battery deployments was an essential input to sound planning. In the first half of the 1950's, after some exploratory flights close to the borders of Russia and getting no reaction from the ground radar stations, photographs were taken of the radar displays providing targeting information for the H2S airborne radar systems.

In 1953, Soviets detonated their first nuclear bomb and followed that by detonating three more. Intelligence on the deployment of these weapons, airfields and missile centres was essential. The USA wanted an aircraft that would fly at an altitude of 70,000ft, undetectable and immune from defence systems. Eventually in 1954 the CIA awarded a contract to Lockheed to build an aircraft based on the F104 fuselage with very high aspect ratio wings of 80ft wingspan that would fly for very long distances at 70,000ft altitude, the U2. Simultaneously, they developed a camera that could photograph and discriminate a football from that height. It was very special in design with a split film, each half being 6000ft long moving in opposite directions, so that the centre of gravity of the aircraft did not change as the photographs were being taken. It produced 4,000 pairs of stereoscopic photographs covering a swathe of country 125 miles wide.

After Khrushchev rejected Eisenhower's proposal for an open skies policy, (which would have allowed frequent over flights of each countries territory by reconnaissance aircraft of the other country), it was decided that the U2 flights would begin. By1956, these flights were being detected by Russian radars but tracking was intermittent and inaccurate.

Russia launched Sputnick 1 -the first satellite- in October 1957. This satellite, in earth orbit, was the first to overfly of much of the earth's surface by the equipment of a single country. (Despite the rejection of the "open" skies policy by Russia, the orbiting of Sputnick 1 marked the start of the programme of satellite reconnaissance by both sides.)

The RAF became part of the U2 programme in the steps towards a joint targeting policy. By 1958-9, the U2 flights were being tracked and vigorous efforts were being made to intercept them with MiG-17 and MiG 19 fighters using the kinetic energy stored by flying at maximum speed and converting the kinetic energy into potential energy in a zoom climb. They were never successful.

By this time the USA was worried about the deployment of Intermediate Range Ballistic Missiles and Inter-Continental Ballistic Missile sites. They were using the U2 flights to try to find them. On May 1st 1960, a U2 mission was due to fly from the south to the north across Russia. The aircraft was suspect in that it had a history of fuel transfer problems that could affect its handling characteristics, particularly when flying near to it's ceiling. There is also some speculation about some engine problem. Whatever the cause, the pilot, Powers was flying somewhat below his normal operating altitude and in a straight line. The Russian defences were well alerted. The Russian SA2 surface to air missile system was a command line-of-sight system that would normally by unstable above about 55.000ft particularly if it was required to manoeuvre. On this occasion, the Russians were probably able, as the U2 was flying in a straight line, to fire the missile as if it were a bullet thus avoiding the need to manoeuvre. Despite having to fire 14 missiles, to bring down the U2, it was a considerable achievement. However, it did indicate that in the future the Soviet high altitude defences would be formidable.

This incident limited the future U2 flights over metropolitan Russia. There was, nevertheless, considerable photographic data to be analysed, jointly by the U.S.A.F and the RAF. It showed, in 1960, that there were very few Russian ICBM sites deployed. However, there remained an uneasy feeling that the Russians were such masters of deception that all sites had not been found.

In parallel with the U2 reconnaissance, a space satellite system was being developed with the ability to photograph the earth's surface from a height of 185 to 270 miles. It was given the code name Key Hole (KH) and had sufficient resolution to be able the detect objects some 2 ft in diameter. Initially, the film was ejected by the satellite, re-entered the atmosphere, and then whilst on a parachute, was caught by a receiving aircraft. It was not long before the digital technology advanced to the stage that allowed the data to be downloaded whilst the satellite was passing over one of the receiving stations sited around the world and located away from Soviet listening stations. The orbit of the satellite defined which targets that could be photographed. It also defined those that could not be visited. Nevertheless, the digital technology enabled mountains of film to be accumulated. Targets of specific interest were always given immediate attention by the photo interpreters. The rest were dealt with by computer comparison of the photographs taken over an interval of time. The computer system could then alert the photo interpreters to the places of interest where changes were taking place.

This information was vital for the Operational Planners selecting the routes to the targets and for the analysts assessing the probability of the aircraft reaching their targets, knowing the deployment of the defences and their effectiveness. In addition, information was vital on the technical characteristics of their radars and missile batteries, such as the frequencies that they were using and those they might switch to in the event of open hostilities. The command and control procedures used when operating these systems were also a key element in deciding how to exploit the weaknesses of their military forces. For example, the Russians controlled their fighters by a "close control" system whereby each separate fighter was guided onto one bomber target. Fighter Command used this system in one-on-one situations, but experience in WW2 had shown that, in multi-multi situations, "broadcast control" was better as it guided the squadron or wing of fighters towards the targets. When our fighters made contact with the enemy forces visually or by radar, they then selected their own targets. The "broadcast control" technique was effective, able to cope with large-scale raids and was much less vulnerable to jamming than the "close control" system. We were always surprised that the Russians persisted with the close control system, opening the possibility that if the communication channel could be jammed, the means of controlling the interception would be lost. Their reasoning may have given priority to knowing exactly where their own fighters were as quite a few of them defected to the West, particularly to Scandinavia. Each defection provided a fund of information on the performance of the aircraft and its equipment. The operating frequencies of the ground radars were also vital so that they could also be jammed to reduce the effectiveness of these defences.

Countermeasures

WW2 provided much experience on the use of countermeasures to assist the penetration of bombers to an enemy target. Many techniques were tried. Spoof and decoy raids, a specialist "jammer" force to interfere with the main radars, use of "chaff" (fine strips of metalled paper to provide false targets) distributed from each bomber to produce large and persistent echoes that both hide the actual bomber responses and confuse the radar operators, communication jammers to inhibit the reception of instructions by the fighters and tail-warning devices to warn of a fighter radar being operated in the search or lock-on mode, indicating a fighter attack. In the late 1950's, the Bomber Command plans envisaged a specialist force of six Valiants for the protection of the rest of the force. This capability was extremely limited because of the narrow frequency band covered and the directional properties of the jamming transmissions. As the main threat to the bombers was going to be surface-to-air and air-to-air missiles, it was evident that each bomber needed its own jamming capability. The combined effects of each bomber's transmissions would, in the right circumstances, be a great bonus.

Each of the Mk2 V-bombers carried a Communications jammer (to counter the fighter control communications), three radar jammers at metric and centimetric wavelengths, (to blank out the radar display tubes to hide the echoes of the bombers), and a tail warning radar (to warn of the search and radar lock-on of fighters). In addition, they carried infrared decoys to deflect infrared air-to-air missiles.

Quite often there were serious paper battles between the Staff at Bomber Command and the Air Staff in London. Sometimes these exchanges involved very hostile letters being sent to and fro between very senior officers. One such exchange involved the purchase of broadband jammers to be used against the surface to air missile systems. Bomber Command desperately needed them to improve the chances of survival on route to and at the target and saw the installation of them on the Mk.2 V-bombers as vital. M.o.D. refused to give sufficiently high priority for them to be put on the equipment list for the contract to be let. At the time, the Mk.2 V bomber cost about £1M and 86 of them were on order. The cost to fully equip all of these aircraft with these broadband jammers was no more than £2M. At one of the Monday morning "prayer" meetings, I suggested that the impasse might be broken if we offered two of the bombers for cancellation and the jammers bought with the money released. It would at least show that we viewed this equipment as the highest possible priority. The suggestion was not welcome. The Bomber Command Staff were determined to win the Whitehall battle. The money would have come from some other part of the RAF budget. Some six months later, we did win and the Electronics Countermeasure Equipment (ECM) bulge was fitted to the Vulcan 2.

To ensure that this equipment would work "on the day", it was tested regularly on the nightly bombing exercises by measuring the frequency band and strength of the transmissions at the Radar Bomb Scoring sites. It was an essential exercise and checked that the frequency had not drifted away from the Russian threat and that the overall reliability of the equipment was satisfactory.

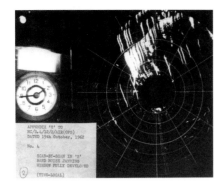

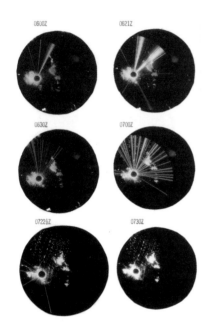

In exercises against Fighter Command, the bomber ECM equipments were tested for effectiveness against real defences with real radar operators and fighter controllers operating with real fighters. One of the Operational Research Branches tasks was to be at the radar control stations to record and understand the overall value that the jamming achieved. On a typical fighter controller radar display screen, there will be a number of permanent echoes, which are shown as bright-ups on the display screen. There is always a cluster around the centre of echoes from terrain and buildings near (up to 5 miles or so) to the radar. At greater range the rest of the bright-ups will be aircraft or, sometimes birds, and any other countermeasures that have been used. It should also be remembered that the operator was able to vary the brightness of the screen (as on a TV) so that he could set it at a level that gives the clearest responses on the display of the targets that he was seeking. Typically, when the chaff is fully developed, the radar responses from the aircraft are heavily masked on the radar screens. The screening is further enhanced when broadband jamming is used.

From these photographs, it can be seen that the radar operator/ fighter controller is presented with a difficult task when he tries to identify a single "bright up" that is a bomber target and then to identify the "bright up" of the fighter that he wishes to vector towards it. The signals from the broadband noise jammers penetrate, not only the main lobe of the radar, but also the side lobes presenting a multi-spoke complex picture on the screen. I have seen fighter controllers turn down the gain on the brightness of the display (to rid themselves of the jamming spokes) so far that they can no longer see any target on the screen.

How the fighter controllers react to these countermeasures depends upon their training and the type of fighter control they were using. Using "close control", the Soviet system, the obscuration of targets makes the task very difficult. Finding the radar response of the bomber and the fighter on the screen so that it can be guided onto the bomber when only one bomber jamming is difficult. With two or more bombers are jamming, it is nearly impossible. Add   communication jamming between controller and fighter and the task becomes virtually impossible.

Fighter Command was using "broadcast control", which got the fighters into the area of the bombers so that the fighters could continue the interceptions using visual or Airborne Radar Interception techniques. By using triangulation of the jamming spokes in the main lobes only from two or three radar stations, it is possible by reducing the brightness of the display tube to a level where only the main lobe jamming spokes can be seen to approximately fix the position of one bomber.

The H2S Navigation & Bombing System. Mk 9.

The Navigation and Bombing System, carried by all of the V bombers, was a key element in their overall capability to reach and successfully attack their allocated targets at night and in bad weather. It was large, complex and had numerous reversionary modes of operation should any part of it become unserviceable. The picture presented to the operator, on the Plan Position Indicator (P.P.I.), is a blotchy set of bright-ups that represent the strength of the reflections from buildings, other man-made features and natural line features. These Plan Position Indicator presentations require considerable skill in interpretation.

If the target was in a position where the presentation was unclear (or an enemy jammer was disrupting the target area of the presentation), then a number of offsets responses (clearly identified locations up to 20 miles or more from the target) could be used to fix the position of the target and bomb it in the normal way, but using an offset for aiming. The nightly exercises against the Radar Bomb Scoring system showed that the accuracy was in the 250 to 350 yd circular error probability (C.E.P.) (This means that 50% of the bombs would fall within a circle of this radius and 50% outside it.) As the main task was to deliver nuclear weapons, whose damage radius was considerable, this was an adequate accuracy.

As the Warsaw Pact high level SAM defences multiplied and were widely deployed, the Command converted to low-level operations. Before the decision to go low level was made, many studies and trials were undertaken. One of those studies was a joint one with the Bomber Command Development Unit to investigate whether the P.P.I. displays could be adequately interpreted at heights from 500 ft down to 200ft above ground level. Techniques were soon developed whereby fixes were taken at about 40 mile intervals along the track and the navigator (radar) was generally able to identify a new fix soon after a previous fix had been used and had disappeared off the screen.

The H2S displays change considerably with height. Comparisons with a Ordnance Survey map and the presentations at 200, 300 and 500 ft above ground level give some indication of these changes. The aircraft is positioned roughly over March at about "10.30" on the picture.

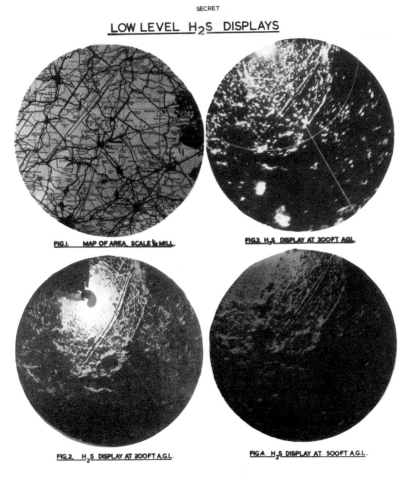

SECRET

LOW LEVEL H₂S DISPLAYS

FIG.I. MAP OF AREA. SCALE ¼ MILL.

FIG.3 H.S DISPLAY AT 300FT AGL.

FIG.2 H₂S DISPLAY AT 300FT A.G.L.

FIG.4. H.S DISPLAY AT 500FT A.G.L.

Ely is shown at the centre of the display. It is a small market town situated in the Cambridgeshire Fenland's - a fairly flat terrain about 70ft above sea level. For the skilled radar navigator, a large isolated factory one nautical mile northeast of the town and a food storage warehouse one nautical mile to the north are the dominant building structures. They gave strong persistent responses at 30 miles range at 500ft above ground level and 19n.m. at 200 ft above ground level. The radar returns shown are from the north-facing side of the town. The southern district of the town was in shadow from the part of the town on higher ground. The twin Bedford rivers were unmistakable from 20 miles range at 500 ft and 15 miles at 200ft. The railway intersecting either of these two rivers provided an accurate fix point. The airfields at Feltwell, Oakington and Mildenhall also provided strong and isolated radar returns. This type if terrain is typical of the western parts of Russia and their western satellites. This trial, with numerous others, was sufficient to convince us all, that from the bombing accuracy and navigation point of view, the conversion to low-level operation could be successfully achieved.

Stand-off Weapons

It had been anticipated that Soviet defences would be continuously improved particularly against the manned bomber. Around Moscow they deployed three rings of high altitude SAM defences. Eventually, the deployment and consequent coverage of their defences would be so extensive that it would be inappropriate to depend upon the manned bomber alone to sustain the deterrent. In 1953, Blue Streak, an ICBM, was proposed as a long-range standoff delivery system. By that time, the essential technologies of guidance and warheads at reduced weight for a given yield were available and the Ballistic Missile Research Agreement with the USA enabled the UK to make a great leap forward missing out many intermediate steps normally required in a development programme. Its development was progressing well. However, doubts were raised about its vulnerability to pre-emptive attack. It was a surface deployed weapon with a slow generation to readiness and, therefore, a poor speed of reaction due to the time required loading the liquid oxygen fuel and readying the missiles for launch.

There were doubts about the effectiveness of such a force if the Command and Control system were required to take action within shorter and shorter warning times. Studies into the cost of putting the missile in silos showed that the price would be very high. In addition, the preparation time to launch from a silo was several hours, making its deterrent potential very dubious. In 1960, the UK abandoned the "fire on warning" strategy and considered that above ground fixed position or silo launched missiles would not survive a first strike. Blue Streak was abandoned that year.

Blue Steel

Blue Steel was an air to surface missile containing a thermo-nuclear warhead and carried on the V-bombers. The development contract was placed in 1956 and was first deployed in Bomber Command in about 1962. When launched at 40,000-ft altitude, it had a range of 100 miles. It climbed to about 80,000ft and reached a Mach 3 before diving on to its target at supersonic speeds. It was propelled by a Hydrogen Peroxide/Kerosene rocket and powered internally by a Hydrogen Peroxide turbine driving a hydraulic pump. The guidance was by an inertial navigation system, with the parent aircraft's navigation system being used to align and correct the missile navigator up to the time of launch. In the early days at RAF Scampton, the custom-built test sets were used to check all the main missile sub-systems, the autopilot, fuel system and the navigator. The complete preparation time for the missile took more than a day. Later test systems checked the serviceability of the complete missile but their diagnostic powers were limited to deciding which, if any, of the main missile systems was at fault. One of the great problems was that all the measurement tolerances were additive so that a unit which had passed one stage might be rejected at a later stage. This situation was undesirable unless some part of the missile had deteriorated. The whole generation process was too lengthy and difficult.

The missile was loaded onto the airfield transmission trolley and taken to the missile test centre. There the missile systems were tested for serviceability and the inertial navigator set up to the ground position of the weapon. From then on the navigation system had to be kept running. The missile on its transporters was then fuelled. The H.T.P. loader/un-loader needed the protection given by his safety suit. The missile is then placed in a blanket and stored under cover until the aircraft is ready to receive it. Readiness was difficult to maintain for long periods and dispersal was almost impossible. However, it did give the bombers the ability to launch outside the missile ring defences around Moscow and other cities and therefore, for a time, did form a useful part of the deterrent. The need to change the launch height and trajectory, when the Command converted to low-level operations, will be covered later.

Skybolt

As the missile threat increased and the prospect that Skybolt would became the next weapon to be carried by the V bombers, trials to demonstrate the Airborne Alert capability of the Command were undertaken.

In the early 1960, the USA asked Britain for facilities to berth Polaris at Holy Loch in Scotland. This proposal became one aspect of a larger package including an American guarantee to provide the RAF with Skybolt. This missile, with a solid propellant rocket, guided by both inertial and stellar navigation, had a range of 1000 n.m. from launch and a megaton nuclear warhead. Such a missile would give the RAF the flexibility that it needed for dispersed and airborne alert operational readiness postures and the capability to stand well outside the range of the Russian surface to air missile threat. The agreement between the US and UK governments was confirmed. The UK had a provisional requirement for 100 missiles. The UK also pressed America to commit to supplying the UK with Polaris submarines, as a measure of insurance should Skybolt development be a failure. This tactic did not succeed and reluctantly the UK had to settle on a gentleman's agreement in this regard.

We were in close touch with the US Strategic Air Command, who had been operating an airborne alert system for some years. They kept a fraction of their aircraft airborne for long periods by in-flight refuelling. In order to validate the Skybolt airborne alert concept, it was necessary for Bomber Command to demonstrate that it could run an airborne alert. We proposed that keeping one V bomber airborne for 12 hours by in-flight refuelling could do this. Using four aircraft, one bomber could be kept airborne continuously. We proposed three trials. The first trial kept one aircraft airborne as long as possible without using in-flight refuelling. Such a trial would give us experience in the reliability of the aircraft, the aircraft servicing load on the stations, the rules for aborting a sortie if the aircraft was not ready for takeoff at the scheduled time and the defect rate from crew boarding through the flight. Whilst airborne, repeated bombing runs every hour against the Bomb Scoring System helped to define the level of serviceability of the H2S system that gave an acceptable performance. The second trial used in-flight refuelling to prolong the flights but the refuelling would take place only in daylight. The third trial required that one aircraft of four be kept airborne 24hr/day.

In reliability terms, the H2S radar system was the most difficult. It had the largest component count of any of the systems on board. The first trial went very much as planned. It gave sufficient experience to go direct to the third trial, particularly as we had realised that at 30,000 ft altitude, where the refuelling took place, the daylight hours were longer than on the ground allowing the flights to be scheduled so that refuelling took place at dawn and dusk at these altitudes. Failures and defects were recorded during the pre-flight checks and throughout the flight. The data showed that the number of defects occurring in the first 30 minutes was 10 times greater than the defects occurring in similar periods for the rest of the flight. The data showed that, if the flight could get successfully airborne, the probability of having to abort the flight was relatively low. The trials successfully demonstrated that the Command could operate using a continuously airborne patrol for part of the force in times of tension, if that were required.

Skybolt was an incredibly challenging system and it was not long before estimates were made of $500m for the development programme. By this time, the UK Navy knew a great deal about the Polaris weapon system. A officer from the R.N. had been working within the Polaris Project Office in the USA for some 2 years. Mountbatten, in particular, wanted the strategic deterrent for the Navy but did not want to fund it out of the Navy budget. The Navy played the "reluctant maiden" role extremely effectively. The "battle of studies" was being fought between the Air Force and Navy Staffs in MoD and the Operational Research Branches of both the RAF and Navy were heavily involved. By this time, the US had well over 30 Polaris submarines deployed using three overseas bases. The system was well established. The RAF studies had a very difficult case to make.

When Kennedy became President, he initiated a review of all U.S. programmes. Skybolt came under particular scrutiny. In November 1962, the US informed UK of the impending cancellation. The UK discussed with them the serious repercussions for us should Skybolt be cancelled. There was an urgent need to consider alternative means of providing a deterrent in the UK with the same degree of independence as Skybolt had.

McNamara offered three alternatives: -
i) continue Skybolt alone
ii) accept Hound Dog , a long-range cruise missile,
or   iii) an alternate system such as Minuteman or Polaris.

To continue Skybolt alone was an unsatisfactory option. It was uneconomic and very risky. Hound Dog did not fit under the Mk2 V bombers and, at best, only one could be carried. To give the UK deterrent the best chance of survival against surprise attack, it must be maintained on a mobile platform capable of long periods on patrol. MOD studied the relative costs and recommended that if Skybolt was cancelled then Polaris fitted with British warheads was the next choice.

Equipment Health Monitoring

At about this time, there was pressure from the Engineering Staff in the Command to adopt equipment health monitoring techniques to improve our understanding of the equipment failure characteristics that could be measured and would show the degradation of some parts of the equipment. With this information, it might be possible to improve the servicing procedures and reduce cost. Strategic Air Command operating a part-force continuous airborne alert were concentrating their flying on a few aircraft rather than distributing it across the force. In that way, they were building up aircraft hours at a considerable rate. The B52 had eight engines and, if one engine fails, it has almost no impact on the safety of the aircraft and not much on its operational capability. The USAF decided to monitor the engine characteristics, such as engine vibration, bearing temperatures, oil usage etc, and to run the engines until they failed, or had to be shut down, rather than have them serviced on the usual schedules. This policy carried the "if it ain't broke, don't fix it" philosophy to the limit. As the engines seldom failed, the effect was to extend engine life at an incredible rate. It had a dramatic effect on servicing policies on both military and civil engines. The information gained from the engine monitoring, enabled engine failure to be anticipated and appropriate action taken in the air to ensure maximum safety and appropriate servicing, repair and replacement action to be taken on the ground.

The Engineering Staff at Bomber Command proposed that the Command introduce Equipment Health Monitoring to the V bombers as quickly as possible. Much to their annoyance, I suggested that they wrote a paper on their proposals for monitoring. In it they should specify how the information gathered would be used to improve both the availability of aircraft and the servicing policy in the Command. It would also be helpful if the paper included proposals on a complex piece of equipment, like H2S, on medium equipment, like an engine, and a simple piece of equipment, like the under-carriage. My view was that there were difficult and different problems with each of these types of equipment. H2S was both complex and operating with old technology. Introducing monitoring equipment involved the risk of causing defects, and reducing the aircraft availability. Their study revealed that no sensible proposals could be made on these equipments.

For the engines, the Command utilisation of its aircraft was only some 350 hours per year. This rate is enough to keep the crews at a high level of operational efficiency but not enough to really explore health monitoring and servicing policies. About once a year, the engines were removed from the aircraft to bring them up to date on the latest modification standard and, of course, at the same time inspect and service them as necessary. The failure rate of the engines at this low utilisation and high servicing rate was negligible. To make engine health monitoring worthwhile, two changes in the operating policy of the Command would need to be made. Firstly, the Command would need to be operating on an airborne alert programme and, secondly, it would need to concentrate the utilisation on a few aircraft.

For the under-carriage mechanisms, no satisfactory proposal could be made. (Health monitoring and self-test was built in to the next generation of equipments. It really seemed to be a case of being just too early with the proposals.)

Penetration Studies and conversion to Low Level Operations

At the beginning of this chapter, I described the history of penetration studies in WW2 and in the 1950's part of the Cold War. In the 1960's, the strike plans were co-ordinated and then integrated with Strategic Air Command. This plan was reviewed on an annual basis. Bomber Command targets included cities, airfields and Soviet Air Defence Systems to be attacked by the combined strength of the Thor missiles and the bomber force.

To do the calculations, on the force required to achieve the successful attack of these targets, involved the study of the targets and their local defences, the other distributed defences and the fighter defences, the alternative routes to the target and possible over-target conflicts with our own and US weapons. After studying the intelligence on the likely mode of operation of the enemy Command and Control System, the effectiveness of the fighters, probable losses to the SAM systems and the effect of our own tactics and countermeasures on their effectiveness, it is possible to make estimates taking both optimistic and pessimistic assumptions of the probability of achieving the level of destruction required against the allocated targets. These calculations were programmed into an Elliot 603 computer. With the analysis of each plan, the estimated success rates of our attack plans fell as the Soviet defence systems continued to be deployed, system over system and layer upon layer.

To fill the credibility gap before Polaris became available, some changes in the Bomber Command concepts of operation were required. The advantages of "going low" were fairly well known. The very large coverage of the medium and high altitude SAM systems were reduced to almost zero. Low altitude SAM systems would need to be deployed in large numbers and be alerted to be effective. The fighters also had a much more difficult task against a low level attack. The Russian system of close control of the fighters would be ineffective. Putting fighters on standing patrol considerably reduced the density with which they could be deployed. Thus, if the attacks were at low altitude, the probability of successful penetration to the target and attack would increase considerably.

The disadvantages of "going low" were also clear. The range of the aircraft would be reduced limiting the target coverage, the training required at low-level would use up the fatigue life of the aircraft very rapidly, the H2S system needed to be effective at low level (described above in the section on H2S), new tactics were required on the use of the jamming systems as at low level more information might be given away to the enemy by having them transmitting than not. Last but not least, the weapons, Yellow Sun Mk2 and Blue Steel, were not designed for release from low level.

These problems were each solved. The range limitations were resolved by reallocating the targets between the Thor and the V-bombers- the fatigue life extended by limiting the training time at low level, and - the H2S capability by the demonstration at the Bomber Command Development of the capability at low level and - a complete review was made of the tactics of using the communication and radar jammers.

The most difficult problem by far was the need to modify the weapons. In order that the fusing and arming sequence of the Yellow Sun, even in a modified form, could be completed and the aircraft escape successfully from the radiation and blast waves from its own weapon, the aircraft were required to pull-up over the target to an altitude of 25,000 ft. The weapon was released at that altitude and the aircraft then dived to low level, turning through 120 degrees to achieve its escape from the radiation and blast from it's own weapon. There were many discussions between myself and the fusing and arming design group at the RAE to ensure that proper account was taken of the relative risks to the aircraft and crew between those threats from enemy action and the risks of damage from ones own weapon. Notionally the risks to the aircraft from the ring defences of SAM sites could be high during the climb. However, these SAM sites, in three concentric rings around Moscow, operated facing outwards from the city. The climb would normally be behind them and, combined with the limited time available to the SAMs to react, the effectiveness of these sites would be seriously limited.

Blue Steel had to be modified before it was capable of release from low level. The pre-planned trajectory was modified so that it climbed to 25,000ft then pitched over into a dive onto the target. The range was reduced from about 100 miles to about 30 miles.

In 1964, the Times Defence Correspondent began to publish articles doubting whether the V bombers could get through. The articles could not go unanswered as they contained numerous incorrect assumptions and particularly, used diagrams showing the considerable SAM system coverage at high altitude that was formidable, but which would not apply at low level. A number of briefings were given by Bomber Command to Chapman Pincher of the Daily Express, famous for his articles on spies, so that he could write in the Express putting the counter arguments into the public domain. He wrote some impressive articles.

The Bomber Command briefings, whether highly classified or unclassified were always impressive. The visitors were taken through a variety of security systems, to the Operational Briefing Room two floors below ground level in the Command Centre. The seats were arranged in rows in front of a very large map of Europe and Russia extending east to the Ural Mountains. The maps were some 9 ft high and 15 ft wide. In front and to the right of this main map were about 10 other maps all hidden and on runners. The lights were put out and the whole room lit by ultra violet (UV) light only. The maps were inscribed and marked with UV responsive paints. The visual effects were very striking. The maps could be pushed across in front of the visitors making a sound that resembled that of a train coming into the station and hitting the buffers. I did not meet a visitor who was not impressed by both the atmosphere and the briefings.

Aircraft Servicing Trials

The Research Branch was involved in many trials in the Command both at the stations and at the R & D ranges. One of the interesting proposals made by the Engineering Branch at HQ was that the Command should change the system of servicing from one which was squadron based to one which used centralised facilities for undertaking the major servicing for the whole station. The principle arguments were that using some degree of specialisation could use the servicing staff more effectively. We suggested that a trial be undertaken at a station, which would give some guidance on the relative merits of the two systems, and exercise the central servicing schedules before they were more broadly introduced. This proposal was agreed. The Station chosen was RAF Wittering. "A" Squadron would use the centralised schedules and "B" Squadron would remain on Squadron schedules. A system of reporting was devised so that the input and output of the two squadrons could be compared. After a three months trial, it was clear that "A" Squadron were operating the more efficiently. The Engineering Branch proposed that the Command should convert as soon as possible to the Centralised system. However, we proposed that before a final decision was made, that the activities of the two squadrons are reversed and the three months trial repeated. This was agreed and a further trial was completed and analysed. Again "A" Squadron was by far the most effective despite using the squadron servicing schedules. It was quite clear that "A" Squadron was the best squadron whichever system was used. In the end, the broad arguments in favour of the centralised system won the day but it demonstrated once again the importance of having some sort of control group to help to understand the results of a trial.

Other Activities

At the time that I joined Bomber Command, I was seconded from the Ministry of Aviation to the RAF Scientific Staff. Each of the Services had their own scientific staff working both in London and out in the field. They were separate with a relatively small number of staff. Their career development was confined to operational research activities, limiting the chances of promotion. This lack of appropriate research and development experience in comparison with those in the larger organisation of the Ministry of Aviation was also a disadvantage. There was reluctance on the part of the Ministry of Aviation staff management to take the RAF scientific staff under their umbrella as they felt that there was a larger proportion of " black sheep" within the numbers, increasing an already difficult problem of integration. I was given the task of examining all the RAF scientist staff records and interviewing many of them in order to write a report on the position. After some months, I was able to report on the study and make recommendations for the integration of the Operational Research staff working for the RAF into the Ministry of Aviation and the steps required to facilitate this process. This integration worked to everyone's benefit.

The job of Scientific Adviser to C-in-C Bomber Command was a great experience for me. I had travelled the USA with the C-in-C and other Staff Officers, witnessed one of the Thor missiles launched at Vandenburg AFB by a Bomber Command Crew, had detailed discussions with the Operational Analysis teams in the Pentagon and at SAC HQ at Omaha and visited the Rand Corporation to discuss the work they were doing in relation to the gradually changing strategic posture of the NATO alliance. I travelled to Australia to see Blue Steel launched at Woomera and discuss, with the UK team there, the process and problems in the preparation of the missile for carriage and launch.

I had the highest regard for the whole Command and particularly the HQ staff. The continuous work with them also revealed their processes of thinking. It gave me great confidence in handling the Assessment and Operational Analysis work that was to follow.

By 1964, I had been in the Scientific Advisers department for almost 4 years. I was thinking "where next?" and my old boss in Aero Flight at Farnborough, Joe Lyons, and now Head of Weapons Department in RAE wanted me back in his team. I was boarded for both the Head of Mathematical Assessment Division and the Ballistic Missile Research Division responsible for developing and firing Black Night, a research vehicle for investigating ballistic missile and re-entry problems. I was selected for the former task and looked forward to a new challenge controlling a Division full of mathematicians who undoubtedly knew much more mathematics than I.

Chapter 5

Mathematical Assessment Division. Weapons Department. RAE

The transfer to Weapons Department was not without its difficulties. The distance to Farnborough was 40 miles, too far for reasonable commuting. Very kindly, Ted and Liz Broadbent invited me to stay with them some nights during the week. It was the year before "O" levels for Tricia and for Vivien, the 11-year-old hurdle. Therefore, it was desirable for them to continue at school for a further year. This period would give time for us to find a house and school for their continued education.

Unfortunately, because the new house was being built and could not yet be lived in, no assistance with the bridging loan was forthcoming from Farnborough. Having arranged a bridging loan with the bank, when I rang some time later to tell them that I was about to write the first cheque, they told me that there was a financial crisis and bridging loans were no longer available. However, after a long and reasoned argument with the bank manager, we managed to reverse that decision. In the end, it all worked well. We bought the house we wanted and Tricia and Vivien, with the support of the Head Mistress at High Wycombe, achieved places at Farnborough Hill for September 1965.

The task in the Mathematical Assessment Division was to assess the effectiveness and cost of a variety of aircraft, avionic and weapon proposals put forward by RAE, other establishments and industry and to advise on the relative merits of the solutions in a variety of scenarios. (The Performance Division of Aero Department did all the estimates on aircraft performance.) The Division was organised into three groups, Offensive and Defensive Weapons, Fuse/Warhead Matching and Reliability.

Offensive and Defensive Weapons

My first critical problem arose over a project, called BL755, which was a cluster weapon designed to improve the effectiveness of freefall weapons. It was a free-fall anti-tank weapon that had been conceived in the Projects Area of the Department. It was intended to employ Hunting Engineering as the prime contractor and Royal Ordinance as the major manufacturing contractor. The name BL755 was typical of the names for weapons at that time. It consisted of two letters and three numerals chosen entirely at random and bore no relationship to the type of weapon or its purpose.

It was a 550lb weapon with an arming rotor on the front. Its function was to provide a time delay before the casing was jettisoned to allow 147 bomblets to be ejected at a spread of lateral velocities to produce a circular pattern around the bomb with bomblets distributed at approximately constant density within that pattern. Each bomblet was variable geometry, starting about the size of a standard can of fruit. The coronet tail then moved back and opened to give it high drag and stability and the nose was extended so that the pietzo-electric fuse was positioned about 2 1/2 diameters ahead of the front of the warhead. (Later improvements to the weapon involved changing the coronet tail for a drag parachute to increase

BL755

the impact angle of the bomblet thereby increasing its effectiveness against the upper surface armour of the tank.) At the back of the bomblet was another arming rotor providing another time delay before the fuse was armed. All these safety devices were required to ensure the safety of the carrying aircraft from any detonations, premature or otherwise, from its own weapon and yet arm it quickly as the objective was to drop the weapon from as low as 200ft above ground level.

The warhead of each bomblet was special. It was a "shaped charge" warhead. At the front of the warhead was a conical copper liner. When the warhead is initiated from the rear, the detonation wave moves forward at considerable velocity and when it reaches the conical liner it converts the copper into a plasma jet travelling at about 11,000 ft/second. If this jet is correctly formed with no side-velocity or spiral content, it has an incredible penetration capability against rolled homogenous armour (RHA) used at that time to protect tanks. Even with a warhead diameter of only 68mm, it is capable of penetrating about 400mm of Rolled Homogeneous Armour (RHA) and have enough residual energy to cause severe damage within the tank after penetration. The technology of the cone and the bomblet construction is key to its overall capability. To estimate the lethality against the tank (and the other targets), the vulnerable areas have to be modelled and the probability of a strike of a single bomblet calculated for each area. Thereby, the overall vulnerability is estimated.

The bomblet was also given a capability against soft targets like aircraft and standing personnel by using notched wire for the casing that, on detonation, broke up into multiple separate fragments.

The problem arose because approval for the project was about to be sought from the Operational Requirements Committee (O.R.C) followed by submission to the Defence Equipment Procurement Committee (DEPC) for the approval to proceed to development and manufacture. Joe Lyons (Head of Department) discovered that there had been no assessment of the capability of the weapon or comparison of its effectiveness with in-service weapons. (I guess everyone thought that it was such an advance on the free fall or retarded 1000 lb bomb that an assessment was unnecessary.) Unfortunately, without such comparisons and estimates of effectiveness, there was no chance that the programme would be approved.

The normal process of assessment meant that it would take three months to complete. The crisis blew whilst I was on leave for a week. To clarify the position, I rang Don Harper, who was Assistant Chief Scientific Adviser (ACSA) on the Chief Scientists staff, whose task it was to ensure the co-ordination of the scientific inputs to the Operational Requirements and the Defence Procurement Committees, to enquire what the situation was with regard to timing. He told me that they wanted the paper in two weeks to meet the ideal timescales. I decided that a new and very simple approach was needed.

At the time, the only weapons available to the RAF were the 500lb or 1000 lb bombs. Relative to 1000lb bombs, the effectiveness of BL755 had to be compared with them against all three types of target. We put our heads together to find a quick way of doing the necessary comparisons. In the simplest terms, the lethal area of one BL755 relative to the 1000 lb bomb was some 10 times greater against tanks, the design target, some four times better against soft targets such as aircraft and 8 times better against standing personnel. It was a weapon that could be dropped by the simplest and the most sophisticated of aircraft. To assess it in all possible situations would require an almost infinite number of calculations to be made for each different type of aircraft on which it could be carried.

To reduce the task to a size that could be completed within the time scale, I proposed that comparisons be made using only three different weapon-aiming accuracies, from the simplest "ring and bead" sight, through the current aiming capability and with the newest weapon aiming system that might be in service in the next 10 years. Comparisons were also required for single weapons and sticks of four bombs. Luckily, the mathematical models required to do the work were already available and in one week the assessment was completed and sent on to London. On the performance data produced by these assessments, the project was approved and went into service with the RAF as well as being sold to numerous other countries. The Russians and the Americans produced similar weapons. The Americans achieved a greater packing density in their weapon but were unable to achieve either the reliability or the density consistency of BL755. But their weapon was much cheaper and achieved greater overseas sales.

During the development programme, it was found that occasionally whilst in free fall the bomblets collided with each other and detonated when pointing in random directions. However on no occasion was there an incident in which the delivery aircraft was damaged by its own weapon.

As a combined team, we had been able to complete the first task very successfully from all points of view. It was now time to take a step back and look at the way we should operate and interface with the Technology Departments of RAE and other Establishments and with the customers for our work in the Ministry of Aviation and the Services.

The General Approach to the Work. It seemed to me to be evident that the Division could play a much more significant role if a much wider spectrum of experts could be involved in the work. We were very short of staff for the work we had to do and it was also clear that we were not going to get more. Apart from the shortage of people who could do assessment work and who had a specialisation of their own, we needed to involve other experts, but only for the period of the study. These would be loaned from: - the Services who would write the requirements for new equipment, experts from the project teams in London and experts involved in developing the new technologies either in the Research Establishments or from industry or any combination of these who could contribute.

This combination might enable us to identify: -

a) the areas of technology that might, if developed, give the greatest increases in effectiveness. It could also identify areas that would not justify future investment.

b) the effectiveness of weapons in a variety of scenarios so that the services could assess the relative importance in each case, and

c) extract some broad cost and risk comparisons to enable a limited number of options to be studied in depth.

It seemed that the only way ahead was to run these studies as working parties, with the outside experts being brought in for the study on a whole or part-time basis. All of the outside experts had a reason for not wishing to be involved. The technology teams were interested in getting on with their scientific research and development and wanted neither the diversions from the work in their laboratories nor advice on which aspects of their technology would pay the greatest dividends in effectiveness or costs. The Services had often been studying the project for years and had already made up their minds. However, if a working party was set up, none of them dare to stay away. They wanted to ensure that their pet programmes were going to continue to be supported.

If we could only get the right tasks, the rest would flow automatically. In addition to the usual effectiveness studies, we were going to need sensible estimates of costs. Costs were a sensitive subject, treated by the contracts branches and cost estimating branches in London as too difficult and sensitive to allow scientists in the Establishments to be involved. For me, they were an essential element of the cost/effectiveness equation. If we were not given access to them, we would have to invent our own. It was not going to be easy and it had to be successful to be accepted. To get started, we needed a task set by H.Q. that was of immediate interest.

The Buccaneer 2*

We had a request from Director Future Systems (DFS) in the Ministry of Aviation HQ for a comparison of a cost and effectiveness of the possible improvements to the Buccaneer 2 as proposals for a largely new aircraft, the Buccaneer 3, had been turned down by the DEPC on grounds of cost. I decided to call together a working party involving everyone who could make a contribution. They would then have the opportunity to feed in their expertise and views and, see the results of the study of both cost and effectiveness. Of course, they may not like the result but, if they criticised the eventual report, it would be from the basis of knowing the scenarios studied and the methods by which each element had been assessed rather than merely reading the report. Many of those attending would wish to limit the options studied to give greater prominence to their own choice. If they succeeded in limiting the options, the likely outcome would be that the report could be viewed as insufficiently broad and, if a choice is made at all, it could be outside the options considered. In my view, such a result would be a disaster, particularly for us as the assessors.

Costs were a very sensitive issue- full of taboos and sacred cows. HQ were very worried about the possibility that chaps out in the field should have any insight or responsibility in this area as they felt that our results might have an influence on the ultimate contractual negotiations. As a Department, we were sure that the assessment job could not be done properly without all of the appropriate costs being estimated as accurately as possible. For our study, they had to be estimated with sufficient accuracy for valid comparisons with effectiveness estimates. The costs used were not merely the price of the aircraft and associated equipment but include estimates of other relevant costs (slice cost).

The costs relevant to the comparison of effectiveness were the peacetime costs of the number of systems that could be brought to bear in war. With the emphasis on limited war in the scenarios considered, only those forces poised ready for intervention really count. Slice costing is an attempt to estimate that proportion or slice of the peacetime budget devoted explicitly or implicitly to supporting the systems being considered.

Four broad types of cost may be distinguished: -

The first is termed loosely development. It includes not only the firms development but also intramural Ministry of Aviation costs, tooling charges and any similar capital items that are in the nature of an entrance fee to the purchase of Buccaneers. Development charges are regarded as independent of both the numbers of systems produced and the overall life of the system.

The second is the production cost, representing mostly an initial investment. This cost is a capital item independent of the system lifetime but proportional to the numbers produced.

The third cost is the operating cost, which is regarded as proportional to the lifetime of the system and the numbers produced. It represents an estimate of all the costs of operating the Buccaneer from an RN shore station,

The fourth cost is the "slice cost." Clearly the operating of a Buccaneer requires the presence of an aircraft carrier. It is necessary to add a share (or slice) of the cost of the carrier to each Buccaneer system.

Our task was to explore the options over sufficiently wide limits to identify the lower and upper limits of the cost effectiveness curves i.e. at the lower cost level, to establish where there was a significant increase in the slope of the effectiveness/cost curve and at the higher level of cost where the gain in effectiveness becomes small, compared with the increase in cost, and the curve becomes rather flat- a typical "s" shaped curve. Nevertheless, limits on the number of options had to be imposed because the answer is always required tomorrow. Our target was to limit the study length to 3 to 5 months.

The first meeting, attended by a nucleus of my own team, representatives of the Directorate of Naval Warfare DNAW, HQ MoA, and the Technology Departments of RAE and the Radar Research Establishment. I opened the meeting by describing the task we had been given and suggested that we would undertake it by looking at a number of options and comparing their costs and effectiveness. This scheme would give us the opportunity to agree the programme of work and, hopefully, avoid disagreements at the later stages. DNAW, represented by Captain Raymond Lygo (later Sir Raymond Lygo) and Commander Pridon Price, reacted immediately. They said that there was no need for that as the Navy had already decided what it required. He described the Buccaneer 3. As it had already been rejected, it readily formed the upper limit of the study. I gratefully accepted his proposal as the most advanced and expensive option.

In order to show the relative effectiveness and costs of the improvements to the Buccaneer 2, MOD asked that the system be broken down as far as possible. Engineering difficulties prevent a complete range of alternatives being possible. The main alternatives compared, with their costs (the slice cost including the cost of weapons) and some greatly abbreviated comments on effectiveness were: -

a) the Buccaneer 2

—Costs-R &D nil, Price per aircraft- £707k, Slice cost per a/c/year- £1.3m

—The improved reliability relative to the Mk.1 should decrease operating costs for the same flying intensity.

b) a above plus both variants of Martell, (TV and AR.)

—Costs. R & D £1.8m- Price per aircraft £730k- Slice cost per a/c/ year £1.57m.

—Effectiveness. Gives 20 miles standoff capability without improvement in direct attack capability

c) a above plus- new Ferranti Inertial Platform

— new Ferranti Forward looking Radar (ex TSR2)

— new rocket ejector seats

— new P1154 Head-up Display

— new P1154 air data system.

— Costs. R & D £18m- Price per a/c £774k- Slice cost per a/c/year £1.41m.

— Effectiveness. For direct attack by unguided weapons increases the target kill chance by a factor of 2 and in any given defence situation reduces the number of aircraft lost per target killed by a factor of 2.

d) c above plus the Martel anti-radar missile AJ168.

—Costs. R & D £19.8m- Price per a/c £797k- Slice cost per a/c/year £1.58m.

—Effectiveness. Adds the stand off capability in some defence situations.

e) c above plus both variants of Martel -Anti-radar and TV guided.

—Costs. R & D £20.2m- Price per a/c £797k- Slice cost per a/c/year £1.69m.

—Effectiveness. Has the capability of (d) plus a 20 miles standoff capability against a wide range of shipping and army targets.

f) e above plus a optical TV sight.

—Costs. R & D £20.5m- Price per a/c £807k- Slice cost per a/c/year £1.73m

—Effectiveness. Has the capability of (e) with an additional fixing capability for the observer, he has low light TV to give a capability against army targets at night and a training aid for Martel.

g) f above plus high lift development and water injection into the engine.

— Costs. R & D £21.8m- Price per a/c £828k- Slice cost per a/c/year £1.77m

— to boost the power at takeoff. (This was effectively the Buccaneer 3, the Navy's choice.)

Estimating these costs involved an incredible amount of work. The more complex and inclusive any system, the greater doubts on the accuracy of the estimates and the confidence placed in them. Nevertheless, they did give a feel for the through life costs and generate a better understanding of the implications of each option considered. In the end, we had to do our own cost estimating. It was interesting to note that in the final submissions to the Operational Requirements Committee and the Equipment Procurement Committee, MOA estimated costs differed from the estimates in f) by a small percentage with R&D at £25m and price at £1.02m and in b), the unit production cost was £0.725m rather than £730k.

In the assessment of the weapon aiming accuracy and overall kill capability of the system, the navigation/attack system had to be treated as one unit. Even the comparison of the radar systems was complex. The reliability was also important. About 25% of the failures on the Buccaneer 2 were Blue Parrot radar failures and as the reliability of the rest of the aircraft was expected to improve, Blue Parrot failures would, in the future, represent a much higher proportion of overall failures. In addition, if we continued with Blue Parrot forward looking radar for another 15 years, it would require that the spares and maintenance support system would have to continue for that time, keeping a production and spares line of rather ancient electronics in being. Even the supply of valves could become a problem and, by the 1970's, become a serious embarrassment. The new radar was expected to be at least 3 times better in terms of target detection probability.

We looked at the possibility of improving Blue Parrot radar or adopting an Elliot proposal for an X/Q band system which an additional capability of giving terrain avoidance and ground mapping simultaneously. Unfortunately, the period before it could be in-service was estimated to be too long. As one would expect, the Ferranti radar generally was the most effective in search range and resistance to jamming.

The variants of Martel gave a standoff attack capability that was important in its own right.

Unusually, one might even say uniquely, I was invited to attend the meeting of the Operational Requirements Committee when the submissions were taken. The task of that committee was to establish the need for the Operational Requirement by the Service making the case and to hear the comments from the other Services. The Navy was asked to make its case. It was based on the need, using the latest technology, to improve the capability of the Buccaneer in maritime operations. The RAF was asked for its comments. I was greatly surprised to hear the RAF say, with tongue in cheek, " We support the Operational Requirement. If we had the Buccaneer, we would want to improve it too." There is no doubt that the new technology would improve the effectiveness by a factor of two and the addition of stand-off weapons would greatly improve both its effectiveness and survival chance. I was asked to expand on some of the effectiveness and cost analysis.

The Operational Requirement was approved and the submission went on its way to the Equipment Procurement Committee. This is a much more powerful Committee, chaired by Chief Scientist or Chief of Defence Procurement and attended by the Chiefs of the Services, the top administrators and sometimes the Secretary of State for Defence and his Ministers. Here, not only the relative merits of the options considered but the part that the system might play in the overall defence concept and the costs and the funding availability within the defence budget. Option b), the lowest cost improvement, was chosen. The standoff capability certainly gave the aircraft a significant increase in effectiveness by allowing it to attack outside the range of short-range defences at minimal extra cost.

For me, it was very valuable experience. The study had demonstrated that this multi-disciplinary, multi-establishment approach could work at my level provided that the request for the work had come from HQ, the Services would co-operate with the preparation of sensible scenarios within which the options could be assessed and HQ would help with the preparation of the cost comparisons.

SAGW- Fleet Defence

At the time, the defence budget was under considerable pressure. The Wilson government was in the midst of a financial crisis and economies were being sought throughout government departments. A number of aircraft projects were under threat including the TSR2, the HS681, the P1154 and the Concorde. In addition, the future shape of the navy was being re-examined. One of the studies underway when I joined Weapons Department was the Flowers Study with very broad terms of reference -" to examine the future requirements for the air defence of the fleet."

Professor Flowers was chairing the Study that was managed for him by the Director General Naval Weapons Mr. Harry Pout. Weapons Department saw this study as a major opportunity to bring its considerable strength in both technology and assessment to bear on all aspects of fleet defence including Airborne Early Warning and fighters and short, medium and long-range surface-to-air missile requirements. The short-range defences also included guns. Andy Stratton took the lead in the Department and used effort from all of his Divisions plus those of the projects area and from other Establishments. The study was given the code name "Confessor".

In the Weapons Department, design and experimental work on the concept and engineering practicalities of achieving missile control by deflection of the jet thrust of the final nozzle had been underway for some time. The jet deflection was achieved by injecting air through holes at 90 degrees spacing around the nozzle, inducing a shock wave across the nozzle and achieving the jet deflection by the change of flow direction through that shock wave.

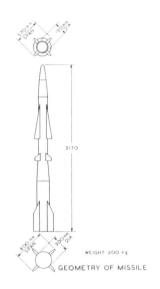

GEOMETRY OF MISSILE
WEIGHT 200 kg

The Sinner Missile

The proposal involved no moving surfaces and presented the possibility of achieving stable flight at *very high incidences*. If such a technology could be used successfully, it would give missiles an incredible manoeuvrability that would be useful in air-to-air combat and enable ship-born missiles to be launched vertically and be turned over quickly to intercept an attacking aircraft or missile at 35metres above sea level and within 100 metres from the ship and countering attacks from any direction. Such a system would also cope with multiple simultaneous attacks from any direction. It could also respond much more effectively against saturation attacks. In order to establish the feasibility of the concept, MoA funded a demonstrator programme of 6 missile launches of a flight programmed but unguided missile. It was built by BAC Bristol and managed by Defensive Research Division Weapons Department under Mike Jarvis. A name had to be found for the programme and it was decided that, as its results would contribute to the "Confessor" studies, logic dictated that before you "confess" you should "sin". The programme was therefore known as "Sinner."

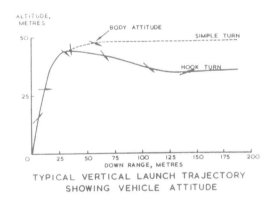

TYPICAL VERTICAL LAUNCH TRAJECTORY SHOWING VEHICLE ATTITUDE

Typical Sinner Vertical Launch Trajectory showing the Missile Attitude.

The Sinner missile, above, was nearly 3.2 meters long. The missile trajectories, right, could be programmed to be either a simple or a hooked turn. The arrows show the attitude of the missile at each position on the flight path. Incidences of over 80 degrees to the flight path were achieved under full control even though the missile was spinning rapidly about its own longitudinal axis. The height/range were measured during the flight and comparisons between the actual flight path and those predicted by simulation were excellent. All six firings were successful.

Normally, on board ship, missiles are stored below decks in the horizontal position. When required for use they are moved onto a conveyer system that delivers them to the launcher to reload it. The process is lengthy, limiting both the firepower of the ship and, because of the other above-deck ship systems, the directions from which an attack could be countered. When stored horizontally, the missiles were much more vulnerable to the acceleration loads caused by explosions on or beneath the ship than if the missiles were stored vertically in their own tubes. The problem was that they either had to be stored on deck where space is at a premium or in tubes with direct exit through the upper deck. Though the latter was preferred, it meant that the spacing between decks where the missiles were stored had to be significantly greater than the current naval practice.

The "Sinner" programme used significant industrial capacity and employed a sizable team at BAC. The capability of the industrial and government teams was notably raised as a result with immediate benefit to other projects. The close integration between the contractors and the R & D Authority contributed to the outstanding reliability achieved in the programme. Many believed that the main reason why vertical launched Seawolf was abandoned was simply the argument that it had never been done before, even though almost all the necessary techniques had been demonstrated. Nevertheless it was viewed as too late

Cartoon by Michael Doone

for the next generation of ship missile systems as the ships had already been partially designed. We were very disappointed. Almost thirty years later, the French used a similar approach in their Aster missile system.

In 1982, when as Director of RAE, I presented a paper at the Financial Times Conference just prior to the SBAC Show, on "The Role of the Research Establishments in the Developing World of Aerospace". As part of that paper, I used the Sinner programme as a fine example of a missile demonstrator programme. Michael Doone, the Aerospace Correspondent to the Financial Times, enjoyed the film and slides presentation, and produced a cartoon of me including the Sinner and the high agility helicopter as the focus of his thoughts.

As mentioned earlier, Joe Lyons policy in the Confessor Studies was to have the whole Department involved as deeply as possible so that we could have a major influence on the final recommendations and any ensuing programmes. Numerous other elements of the work contributed to the overall M.O.D. study. They included: -

i) the best form of guidance for a particular missile attack envelope requirement.

Line of sight SAM systems were cheaper and sufficiently accurate for the shorter ranges, semi-active guidance was needed and was more expensive for the longer ranges and if the overall rate of fire was to be improved with either of the above systems, active missile terminal guidance was required, the most expensive.

ii) for short-range defences, guns seemed to be the natural vehicle but they were insufficiently accurate to hit a small target like a missile. Considerable work was put into investigating techniques of estimating the end-game error between the shell and the incoming missile and then taking advantage of the spin of the shell to eject a mass from it in the opposite direction to the correction required so that the terminal error of the shell could be reduced or eliminated. Even though BAC and ourselves did many studies, we were unable to establish that the technique was cost/effective. (It was a desirable objective even though the programme was called "Bonkers".)

iii) Fighters and a ship-born Airborne Early Warning aircraft for long-range defence. We had just completed the study of ten options for an A.E.W. system when it was decided as a result of another study in London that the A.E.W. system would be land based. This study was part of the future of the "carrier battle" that was to last a number of years.

During the Confessor studies, Professor Flowers wrote a paper, in the tradition of Zuckerman, called "The Squatting Pigeon Alerting System." It was based on the natural habit of pigeons, when on the ground, to keep a sharp lookout for any threat. Once alerted, they turn towards the threat and squat. It follows, therefore, that with a little training to recognise an approaching aircraft as a threat, a magnet fastened to them to determine the direction that they are facing and a micro-switch below them to sense when they squat, a very low cost alerting system could be produced. It was great fun at the time. There were a number of the less scientific that did not, at first, recognise it as a joke!

It was at about this time that Joe Lyons was appointed to be Head of the Road Research Laboratory and John (later Sir John) Charnley took over as Head of Department.

The TSR 2, it's Cancellation and Subsequent Events.

In respect of the TSR 2 and its cancellation, it is difficult to separate the true history from the mythology and the rhetoric. The General Operational Requirement G.O.R. 339 was conceived from thinking of the need to develop an aircraft as a replacement for the Canberra capable of reconnaissance operating in the European theatre. O.R. 343 followed as a result of studies that included operations East of Suez. At the time, there was no requirement for a strategic deterrent or strategic strike role though, with the growing miniaturisation of nuclear weapons, that eventually became feasible, about 1961. This additional capability was then looked upon as a bonus. However, it eventually became the main driver of the requirement. By 1965, the UK was in financial crisis, withdrawing from East of Suez. At the same time the Plowden Report on "The Future of the Aircraft Industry" emphasised that collaboration, particularly in Europe, was the most likely policy.

The cost of the TSR2 was escalating, too many national resources were being deployed in the aircraft industry and, if the future of the industry lay in Europe, it was argued that it was necessary to cancel the TSR2 to release funds for joint programmes. The last straw appeared to me to be the cancellation by the Australians of their prospective purchase in favour of the F111.

In addition, the existence of the American F111 programme also contributed to the cancellation of the TSR 2 programme. The American Government offered Britain the opportunity of purchasing the high performance multi-mission F111 as a low cost substitute for the TSR2. In fact, 50 F111 might well have been only half the price of 50 TSR 2. There was also the argument that the servicing and maintenance costs of the F111 were claimed to be considerably lower. An initial contract for the F111 was placed. However, the inter-service battles in the USA prevented the aircraft from being developed as a multi-role, multi-service aircraft. It also suffered from intake/engine compatibility problems caused by vortices shed, from the fuselage, ahead of the engine intakes and then ingested, resulting in the engines surging. The Americans had been unable to find the cause of the problem. It was a considerable feather in the RAE cap that John Seddon of Aero Department identified the problem and recommended successful modifications for its solution.

After the cancellation of the TSR 2, the RAF was convinced that it must have a variable geometry multi-role aircraft. The F111 could fill that role. Collaboration with Europe was still being sought and Dennis Healey argued in parliament that the F111 and a collaborative European aircraft were complimentary. (The F111 was eventually cancelled in 1967.) In June 1965, the UK and French governments agreed that a study should be made of a variable geometry aircraft, in order that the fundamentals of an Anglo-French project could be agreed by mid 1966. To do this, the following timescale for the studies was required.

July 1965-   *Operational requirement V/SOR1 for the RAF and RN and the French AF and Navy.*

Aug.-Oct 65-   *Studies in industry of the possible solutions to the above O.R.*
The study of the aircraft was given to BAC and GMAD and the engine to SNECMA on 1st Aug. 65.
Studies of the equipment fit were given to Ferranti, Elliots, GEC, HSD, BAC and later EMI.

Nov 3rd-9th-   *Presentation by the firms on their studies was made at RAE and the brochures made available.*

The aircraft all up weights to be studied were 30000, 40000, and 50000lb with special emphasis on 35000lb to suit the French carrier requirement. Five different roles were to be studied including-
i) specialist fighter
ii) fighter aircraft with some strike capability
iii) dual capability fighter/ strike aircraft
iv) strike aircraft with some fighter capability
v) specialist strike aircraft.

In RAE, to facilitate the work involved, four working groups were set up to examine the proposals made by the firms.

The Aircraft	Chairman Mr. H. Plascott
The Radar/Nav./Attack Systems	Chairman Mr. A. Walker
The AAGW	Chairman Mr. W.M. Jarvis
The Overall Weapon System Assessment	Chairman Mr. T H Kerr

These groups were asked to report by 20th Dec. 1965.

In the time available, the firms made an effort to cover the range of possible solutions to the operational requirement and some useful proposals were made. The aircraft firms did not cover the range of all-up-weights required but limited their efforts to a 35,000 lb aircraft and gave some performance changes possible on a 44000 lb aircraft with the same equipment fit. The equipment firms proposed a range of solutions. It was the task of the Weapon Systems Assessment group to examine the variation of effectiveness with the performance of the aircraft and equipment and suggest the most promising solutions by comparing their cost and effectiveness with other systems current or available in the near future. All working groups worked in parallel.

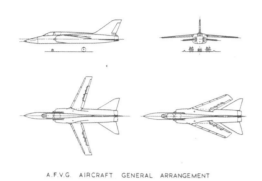

A.F.V.G. AIRCRAFT GENERAL ARRANGEMENT

AFVG Aircraft General Arrangement.

The Anglo-French aircraft design selected for assessment in this study and the typical missiles studied were the new concepts of the day. The number of scenarios and the number of variants to be studied presented us with a massive amount of work to be done in a very short time. The final assessment document was some three inches thick. At the time of reporting the

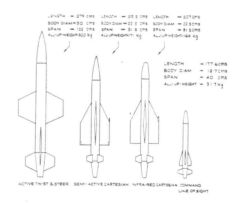

Typical Missiles being considered in the Anglo-French V.G.Studies.

engine position had not been defined but we had assume its performance. We had serious doubts that the 35,000lb aircraft was big enough to carry the electronic and weapon fits specified. In fact, we greatly favoured a 50,000lb aircraft of the size of the Phantom. However, the whole technical specification was being driven by the French carrier limitations.

The French were using "tonnes" as their weight measure (2000 lb) and were targeting 16 tonnes for the all-up-weight. The reports had been produced so quickly that they had not been vetted to the usual standard. To get them out on time, they were produced as Draft with a Weapons Department number and issued widely to H.Q. We had worked incredibly hard, long hours each day and at weekends. Late one night one of my chaps suggested to me that we should put a quotation on the inside title page. It was against my instinct but, as it was only a draft, I agreed. He suggested a quote from an old miner's song " Sixteen tons and where do we get, another day older and deeper in debt." Evidently the tension in London was too great to cope with the risk that the French might see this quote and be offended by it. We received a Top Secret signal with instructions to withdraw the report. With much embarrassment, we reprinted the inside title page and issued it to all addressees with the request to replace it and return the removed page to us. I am sure that the pages were replaced but the offending ones were never returned to us. They must have achieved a value of their own. My lesson was- never to put a "quote" on a report again.

Eventually, the negotiations with the French failed and we went through a number of studies on an aircraft called the UKVG to investigate the possibilities of going it alone.

Ultimately, the UK, Germany, Italy and Spain came together to produce a common specification for the Multi-Role Combat Aircraft, MRCA that eventually became the Tornado Fighter and Strike Aircraft with a more relaxed weight specification. The MRCA was really a political aeroplane. I do not believe that there was ever a formal Operational Requirement for the aircraft nor approved by the MOD Committees in the traditional way.

Fuse-Warhead Matching

Under Geoffrey Pullan, I had a small group of two whose task it was to do the research and project work of the fuse and warhead matching for all UK anti-aircraft missiles. The

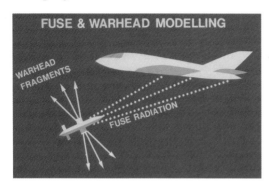

An Aircraft/ Missile Attack End-game.

mathematics of the typical dynamic end-game scenario of a missile attacking an aircraft was impossible to simulate without a complex computer game requiring simulation of the relative motions, the infrared or radar fuse beams and the fragment beam from the warhead. The target passes through the fuse beams that measure the relative speeds, direction and distances of the target and the fuse triggers the warhead to give the maximum chance of damage to the target. The targets could be a wide variety of size and configuration, the fuses could be simple omni-directional capacity fuses, single beam or multi-beam radar or infra-red fuses, and the warheads could be simple blast warheads, fragment warheads, continuous rod warheads or, in the case of the largest and complex missiles, a directional warhead. Each of these changes had a major impact on the size and cost of these items as well as having an important effect on the overall cost/effectiveness of the missile itself.

At first sight, it was surprising to find this type of work with an Assessment Division. For me, it was very welcome. It gave me an opportunity of contributing to some scientific work that was not continuously under the stress of inter-service or inter-establishment arguments.

The team took data from

- Mechanical Engineering Department whose responsibility it was to establish the vulnerability of aircraft to attack by various types of warhead and create a mathematical model that could be used in a computer programme. To obtain the data, they usually took parts of aircraft or missiles that were redundant prototypes i.e. past their useful life and no longer required, and detonating warheads in a variety of positions nearby. The resulting damage was then assessed and converted into lethality data. These experiments enabled a good model of the vulnerability of our own aircraft to be built. These models were then converted to a more generalised model representing likely enemy aircraft. Sometimes we were able to refine them further by intelligence data or even parts of enemy aircraft etc. obtained from other areas of conflict. (Another part of Weapons Department was responsible for supplying the data on missile vulnerability.)

- data from the fuse experiments to give trigger points. The E.M.I. radio range at both Feltham and Wells and any direct experiments largely supplied this information by the fuse designers' own facilities.

- a direct contract from RAE to E.M.I. at Feltham, who designed, maintained and operated a simulator to study aircraft/missile and missile/missile engagements under a wide variety of conditions so that the best overall configuration of the fuse/warhead combination, (recognising the constraints of weight and space), could be chosen. At about this time sufficiently large computers were available (Atlas) to do the work digitally.

The "kill" categories that were in common use in US/UK/Canada collaboration at that time were: -

Cat F (t): the target would become permanently incapable of maintaining directed flight or trajectory within a time interval of t seconds after the damaging strike.

Cat C (t): The target would become unable to continue the stated mission within the time interval of t seconds after the damaging strike.

Cat E (t): the target would receive damage necessitating repairs that would be completed in time T days.

The early anti-aircraft G.W. warheads were usually fragmenting types, fused by fixed-look-angle fuses. It was found that the lethality from fragments was low because the aircraft designers had configured the aircraft such that the parts of the aircraft vulnerable to fragments were generally small, duplicated and widely dispersed. The task of optimising the fragment beam width and the fuse parameters (look angle and time delay) to ensure the requisite strikes on the vulnerable parts was difficult and led to the lethality of the fuse-warhead combination deriving largely from the blast bonus over a wide range of engagement conditions. It appeared that the target was more uniformly vulnerable to blast wave damage and the fuse/warhead matching less critical so a change was made to external blast warheads.

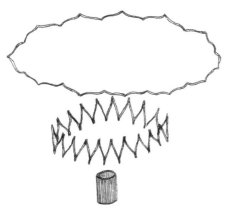

A Continuous Rod Ejection.

Unfortunately, blast warheads suffer from serious degradation of effectiveness with altitude, by as much as a factor of between 2 and 4 at 60,000ft. The alternative warheads include sub-projectiles, shaped charges and continuous rods. The continuous rod (C.R) warhead was a series of steel rods, typically, six inches to 1 ft long laid side by side in a cylinder just within the diameter of the missile and welded together to form a continuous loop. When the warhead was detonated, this cylinder of rods expanded and stretched, like a concertina and remained continuous until it reached its fully extended diameter. The objective of the C.R. warhead was to remove a large piece of the structure of the aircraft whereas the fragmenting warhead achieved its effectiveness by the penetration of fragments into the vulnerable areas of the aircraft. Because of fusing difficulties and lack of reliable vulnerability data, the majority of studies concentrated on the use of fixed look angle fuses with either blast or continuous rod warheads.

The choice varied with altitude, guidance accuracy and warhead weight available in the weapon. Generally it was found that the continuous rod warhead was preferable for attacks at all altitudes up to 60,000 ft where the miss distances were greater than 30 ft root mean square (R.M.S). However, when missiles with warhead weights of less than 100 lbs and root mean square miss distances less than 15 ft. RMS were attacking targets at low altitudes the differences in the warhead design were less well marked. The final choice requires careful analysis and may depend upon which damage criteria has been specified for lethality. Wings were also very vulnerable to continuous rod warheads. Against an air to surface missile, the continuous rod was also superior.

The relative merits if single beam and multi-beam fuses were also compared. Broadly it was found that the effectiveness of a three-beam fuse was not improved significantly by adding further beams.

We started studies on the possibilities of improving warhead and weapon effectiveness by directing the damaging energy at the target using an aimed warhead. The doubts on the effectiveness of the shaped charge were resolved by experiments using the high-speed track at Pendine. The merits of high ejection speeds (many times those of fragments) that might ease the fusing problem, and give considerable penetration capability with associated post-penetration vapour damage had to balanced with the extra missile size needed to carry such a warhead.

It was particularly rewarding that we were able to get the increased funding that enabled Geoffrey Pullan to considerably increase the pace of the work to keep up with the increasing demands for it.

Reliability

Reg Chaplin ran a small team that attempted to predict the reliability of missile systems and their major elements by looking at the laboratory tests of the reliability of the separate components tested under appropriate environmental conditions and then calculating the likely reliability of the whole system in operation. It is an extremely difficult task. The work was being done in an atmosphere of considerable scepticism that such predictions were possible, particularly as most of the results predicted an overall reliability much lower than that required. The controversy had the very desirable effect of increasing the understanding of reliability and greatly increased the resources applied to environmental testing. (When I returned, in 1972 to take over the Weapons Research Group, the Reliability team had achieved an excellent reputation. The Concorde Project team were being pressed by the UK and French companies to fund the redesign of the engine intake and nozzle controls to achieve better reliability. Chaplin and his team applied their techniques to the problem and strongly recommended the system design be changed from analogue to digital, as this was the only way that the required reliability could be achieved. The funding was approved).

By this time, the conversion of the Army Operational Research Establishment to the Defence Operational Research Establishment at West Byfleet had been working in this new mode for about six months. I was promoted to become a Deputy to Terry Price, the new Director, to take responsibility for the Land, Sea and Air Divisions of the Establishment.

Chapter 6

The Defence Operational Analysis Establishment

In 1961, Robert McNamara became Secretary of Defence in the USA and stressed the need for a sound planning-programming-budgeting system for managing in defence. Its objective was the most effective allocation of resources. His philosophy drew heavily on the operational analysis in WW2. He was assisted by Charles Hitch, Assistant Secretary of Defence and Alan Enthoven, Assistant Secretary of Defence (Defence Analysis). The USA Defence Department was supported by considerable operational analysis strength both in government and independent groups such as the Rand Corporation in California.

Dennis Healey was Secretary of Defence in the UK and was greatly impressed by the potential help that he might get from similar support in the UK. DOAE was to be the UK tri-service equivalent operational analysis group, all in-house. I joined DOAE at West Byfleet when it had been in existence for about six months. The staffs were a mix of scientists, serving officers and administrators from the naval, army and air environments. The essential elements to the success of such an arrangement were: -

i) getting tasks that is relevant, of considerable importance and that can be addressed and answered in a period of less than nine months. (If the work takes longer than that, the management will have lost interest or moved on.

ii) getting suitable scenarios agreed by the three services that are relevant to the examination of the question to be answered.

iii) being allowed the freedom to look at the sensitivity of the analysis to change of parameters where such changes make a considerable difference to the results.

By kind permission of the Evening Standard.

Starting points such as these look sensible and easy to obtain *but they were not*. Air Marshal H.N.G. (Nebby) Wheeler, (later Air Chief Marshall Sir Neil Wheeler) who was Deputy Chief of Defence Staff and Chairman of the Operational Studies Committee, was a great help to me in gaining the co-operation of the three services in this vital planning stage and in many other respects. The work often involved considerable emotion. Everyone involved had strong feelings about it. It could affect their lives, their livelihoods and, in the limit, all our precious freedoms. Many people hated the application of dispassionate analysis to warfare and strategy. A very amusing cartoon, above, appeared in the Evening Standard, on 2nd April 1968, expressing this sentiment. The caption read, "Doubtless some of you guys are asking yourselves what sort of position LBJ has left us in Vietnam now."

However, there was no sensible alternative in a nuclear age when technology could offer many different choices to solve a defence problem. We had to try to contribute to the decision-making process and offer support to the political and military judgement that had to be brought to bear to identify the best solutions that matched the national objectives at that time.

To make a useful input, the objectives had to be identified, the method of measuring effectiveness agreed and the options costed in a way that reflected the reality of the defence situation, i.e. full, opportunity or marginal costing applied as appropriate.

DOAE

The Army Operational Research Establishment (AORE) had been in existence for many years, working exclusively for the Army Department. It was sited in a lovely house, previously owned by Worthington brewing family, in West Byfleet. In the gardens of the house was a hutted encampment where the bulk of the staff, the war gaming rooms and the computer were housed. The work of AORE assisted the Army Department to understand the complications of battlefield scenarios, the interaction between weapon systems and their influence on tactics. In addition, it had a particularly important role in giving practice to military commanders in the command of troops in the field in simulated battle conditions. AORE also undertook field trials of various equipments. The Army felt that, with the formation of DOAE, it would loose its support from that Establishment.

That gave DOAE a good reason for separately including in its programme about 30% of the work for all three separate Service Departments, the bulk (about 70%) being joint service studies. I also believed that it was important that the Central and Service staffs in MoD should be a high regard for the work. Multi-service studies often pleased few because no one quite got the recommendations what they wanted. The single service studies were a very important part of the work at DOAE and helped to maintain good relationships with the single Service Departments.

The Mathematical Assessment Division in RAE, concentrated on single weapons or aircraft, mostly in one-on-one or one-on-many engagements. DOAE looked at a higher level of military operation involving combined units, prehaps from all three services and was often required to assess the need for any particular equipment or combination of equipments in a single or combined service operation and the size of the force from any of the services needed to carry through the defined operation.

DOAE had two hierarchical routes to the Secretary of State. The first was through the Assistant Chief Scientific Advisor(Studies and Nuclear). The second was through, Special working parties at ** star and *** star level, whose task was to set the operational scenarios to be studied, the Operational Requirements Committee, Weapons Developments Committee and Defence Research Committees through to PUS and the Chief of the Defence Staff. I was always grateful to the Deputy Chief of Defence Staff for his help in getting studies that mattered, getting the scenarios agreed and a timescale for the work that gave us a sporting chance of producing results that really had an influence on defence thinking.

The Defence Operational Analysis Organisation in MOD included Operational Analysis working with each Service not only to support the case for the proposals of that particular service but also to produce arguments that rebut the proposals of the other Services.

When I arrived at DOAE, I took over a new post AD/Ops with the Land, Sea and Air Divisions working under me. Luckily I was able to take with me from Farnborough, Bert Longdon and John Shrimplin. Hal Hood, previously with me at Bomber Command, was already the Superintendent of the Air Division.

Some typical studies involved Support Helicopter Requirements, Limited War Operations, Nuclear War in Europe, Communications, Command and Control, Maritime Operations, Reinforcement Studies, Conventional War in Europe, Anti-Submarine Warfare, Night Fighting, Helicopters v Tanks, NATO Air Defence, etc.

The SAM/Fighter Working Party 1966/67

The terms of reference were "to study, on a cost/effectiveness basis, the need for a new medium range SAM system to meet GAST1210 to operate in conjunction with the Phantom/Sparrow interceptor and other SAM systems available in the same time scale (1975-1985) and to suggest a broad outline of a ground radar system (AST1514). The scenarios considered were, protection of an airhead, protection of the army in limited war and protection of the army in Europe. A draft report was required in six months and a final report in nine months. The underlying question was: - "With what, if anything, should we replace the ageing Bloodhound and Thunderbird systems?" A number of generic systems called PX430 were possible. How do we make the choice? A typical

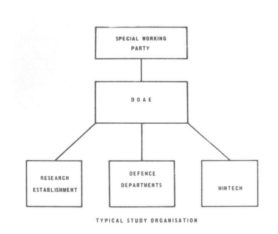

TheOrganisation for Multi-Service Studies.

working party organisation is shown on the left. The members of the working party included representatives from all three Service Departments, the Research Establishments, MINTECH HQ, and a significant scientist and military team from DOAE. I chaired it.

In the early 1960's, the USA was proposing a short/medium range SAM system called Mauler. Its speed of reaction and radar capability were such that none of the systems proposed in the UK matched it in price or in capability. We had grave doubts about these cost and performance claims and it was not until the early development trials that the real reaction times and costs were revealed leading to its cancellation.. RRE Malvern had responsibility for SAM systems (except for the missile which was an RAE responsibility). Two of their scientists, Colin Baron and John Twinn considered how a short range system, could be put together to meet this urgent defence need. They concluded that an optically guided, command line-of-sight system with the weapon being a "hitile" i.e. the missile is to hit the target, penetrate it and after a small delay, explode the warhead within it. The optically guided system was called ET316 and the radar blind fire version was DN181. They became Rapier and Blind-Fire Rapier. The optical guidance was very resistant to jamming and decoys and it could be produced at a price that was affordable.

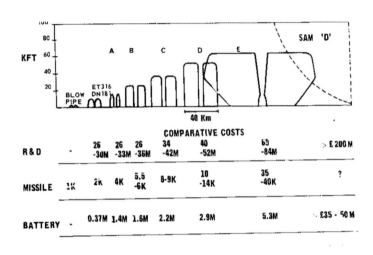

Spectrum of Systems considered in the SAM/Fighter Working Party.

We considered a spectrum of SAM systems, with their R & D, missile and battery costs. Blowpipe was a Shorts designed shoulder launched weapon-provided for platoon air defence. ET316 etc. are described above and to be deployed in the field by the army for HQ and brigade defence. The possible PX430 systems are shown as systems A, B, C, and D. These systems could have vertical missile launch. System E, was Land

Dart, a land version of the naval system Sea Dart. The American system, SAM D, was the early proposal that eventually was put into service as Patriot. The diagram also gives our estimates of the R & D, missile and battery costs. Each of the new systems could have an anti-aircraft and anti-missile capability. The offensive and defensive systems, below, included all relevant systems.

System E. Land Dart.

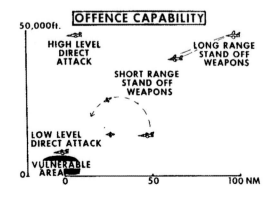

Offence Systems Considered.

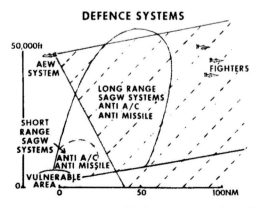

Defence Systems Considered

We always emphasised that we assumed that the offence objectives would be to surprise or evade the defences and exploit its weaknesses. If these objectives failed, then the offence would try to saturate the defence systems. On the other hand, the defences would always try to make all offensive options equally unattractive. The latter is the much more difficult task. Many different computer models were used to study the separate elements of this interactive war game. They were then brought together manually to enable us to reach our conclusions.

During the progress of the work on the defence of airfields and the defence of a task force at sea, two proposals caused heated discussion. The Rapier deployment for the army, either in H.Q. defence or on the field assumed that considerable mobility was required and that the time required to deploy allowed only for the locating of the systems looking outwards over one sector from the area to be defended, the site being chosen to give minimum screening by terrain and trees etc. in the direction to be defended. The systems would therefore be operating independently, particularly with regard to electro-magnetic transmissions. Airfields, on the other hand, are static targets. If the Rapier sites could be elevated by placing on top of artificial mounds or even placed on the roof of hangers, then the firepower of all the systems could be brought to bear against an incoming raid.

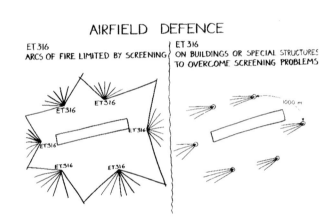

Rapier (ET316) in the Defence of an Airfield.

Against a raid of 10 aircraft, eight elevated Rapier systems would provide a defence equal to that achieved by 16 Rapier systems deployed at ground level. RRE had made something of an industry of examining the screening at many army sites in Germany where Rapier might be deployed, but their involvement in this work did not explain their reluctance to accept that, for the defence of airfields, there would be an advantage in elevating the missile sites. I ultimately discovered that they knew that the command guidance of the missiles from different fire units would interfere with each other if there were direct lines of sight between them. It took some time and a lot of work to get the electronics of Rapier modified to avoid this problem.

It is difficult to summarise the results of such a complicated study. Three strengths of attack, six levels of defence and two levels of expected damage were compared. The overall defence cost at these damage levels of providing the different defence mixes over 10 years in peacetime for two airfields were compared. The report ran to three volumes and was completed within the specified time. There was a large step increase in cost and effectiveness when AEW and fighters were included in the mix. If attrition was to be achieved against aircraft carrying stand-off weapons, AEW and fighters were essential.

The conclusions reported were: -

i) the principal threat is considered to be cruise missiles with H.E. warheads, mainly air launched and subsonic strike aircraft at low level and at heights up to 50,000 ft.

ii) the raid strengths, against the airfield, were realistic and yet considerable defences are required because relatively simple future strike systems can inflict considerable damage,

iii) the army in the field in limited war presents area and point targets which may be attacked from the air. A SAM system of C, D or E capabilities is essential to ensure that defences are available against such attack options.

We find, therefore, that we needed a SAM system, to meet GAST 1210, against the standoff missile and strike aircraft threats. Of the solutions considered, and bearing in mind the cost of each solution:

System C is the cheapest system that approaches a solution to this requirement. A solution of lesser performance does not seem to provide a worthwhile addition to our defence armoury.
System D comes nearest to meeting the requirement and counters the bulk of the threat.
System E has significantly more range and height than the other systems, but its capability is, perforce, inefficiently used in the scenarios considered. It is also the most expensive.

All in all, we felt that it was a considerable achievement to get agreement on the outcome of the study and report on time. It had exposed an important problem in Rapier, which, when corrected, facilitated much greater ease of deployment and increase in its effectiveness. The study had also given good operational and cost reasons for limiting the effectiveness envelope of medium SAM solutions. The output of the study must have been welcome in MOD as we were soon to be launched on a series of studies on operations on the Central Front in Europe.

Studies of Future (post 1975) Air Operations in Europe

Early in 1968, it was decided, by the Future Combat Aircraft Steering Group, that there should be a thorough study of the future air operations on the Central Front in Europe. Four studies were set up: -

Study A- the analysis of strike and reconnaissance operations- chaired by Jock Henderson, Chief Scientist RAF.

Study B- the analysis of NATO air defence post 1975- chaired by me.

Study C-satellite reconnaissance as an alternative for battlefield intelligence.

Study D-aircraft survivability on and over the battlefield.

The terms of reference of Study B were: -

i) to assess the nature and level of air defences required to counter the enemy threat to the shield forces of NATO post 1975,

ii) Specifically the study is to-

a) take account of the air defence systems already projected for NATO in this period and the likely changes in NATO's air defences both over and to the rear of the potential battle areas which might be feasible as a result of technical forecast developments;

b) establish whether there are requirements for a fighter aircraft and surface-to-air missiles, together with their ancillary systems such as air-to-air weapons, radar, AEW, control and other relevant aspects. The best systems mix should be indicated, together with the likely efficiency of the whole environment against enemy attack.

The working party included representatives of all three Services, MoA HQ, the Research Establishments and representatives from the air and land divisions of DOAE.

The Concept of NATO Operations and its Consequences for Air Defence.

In NATO Defence Policy, there had been a recent shift away from the rapid recourse to nuclear weapons towards the desirability of maintaining, for as long as possible, a defence by conventional forces alone. A consequence of this change was that NATO's air defences needed to be capable of offering protection not only to forward units engaged in battle but also over the rear combat areas in which the air and logistic support were deployed. Thus NATO's air defences- both in the central region and on the flanks- should give cover two to three hundred kilometres back from the border to protect the majority of these important rear areas.

A factor of major importance in determining the size and shape of an air defence system is the size and shape of the threat that it would have to counter. Intelligence forecasts of enemy orders of battle do not normally attempt to look more than three to five years into the future. This study required a forecast to about 1980. Intelligence, combined with scientific and military judgement, produced a likely threat of-

-Attack aircraft like FITTERs or FISHBEDs, FLAGONs and FOXBATs.
-Reconnaissance aircraft including FISHBEDs, FLAGONs and FOXBATs.
-Air Defence Aircraft including FISHPOTs, FLAGONs and FOXBATs.

Typically, the Warsaw Pact air threat in 1980, opposite 2 ATAF, could be several hundred attack/recce aircraft, plus air defence fighters employed in a secondary role. We also assumed that there would be no cross reinforcement between sectors.

NATO's 1967 air defences- the ground environment NADGE was still incomplete after eight years of work and suffered from dangerous weaknesses. It was almost totally ineffective against low-altitude attack. The ground forces in forward areas had little air protection other than air defence guns, the kill potential of which, against manoeuvring targets, was low.

However, they might have some value in denying enemy aircraft complete freedom to engage in undisturbed attack. Existing medium altitude missile systems had no significant capability against low altitude attack. Although NATO interceptors had some capability when organised for vital area defence in the low-level visual interception role, they had only low effectiveness against wide ranging low level attacks.

The Warsaw Pact air commanders must have been well aware of this deficiency in the NATO air cover and there was intelligence evidence of their interest in low-level attack tactics. Thus, the prime requirements for the development of NATO's air defences were clear—protect the forward and rear battle areas, remove the existing weakness at low level and be capable of dealing with a threat that included large numbers of aircraft of modest performance and a smaller number of advanced capabilities.

Components of a Future Air Defence System

In terms of development timescales, 1980 was not far away. Most of what might be available was already in the feasibility stage or under development. The aircraft considered were: -

i) Phantom F4K. A modern interceptor carrying 4 Sidewinder and 4 Sparrow air-to-air missiles. Cost £1.5m each

ii) Improved Jaguar-with a major new avionics fit and supersonic capability. £1.2m each

iii) Multi-Role Combat Aircraft (variable geometry) carrying 4 Advanced Air-to-air missiles. £1.8m each.

iv) FX type aircraft of high-energy manoeuvrability. £2.1m each.

v) V/STOL strike aircraft. £2.5m each.

vi) F111 type. £2.1m each

vii) YF12. capable of M=3 and 100,000ft . £5m each

The Surface-to-Air Missile Options.

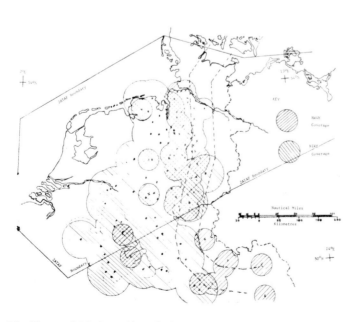

The Nominal Medium Altitude SAM Cover-1968.

In 1968, the missiles then deployed by NATO were Hawk, with a range of around 25kms and a ceiling of around 50,000ft against Mach 1 aircraft, and NIKE with a range around 50km and a ceiling around 80,000 ft. Some Nike missiles had nuclear warheads. Neither system had any significant capability against low-level targets.

The future S.A.M.'s that were possible were: -

i) Rapier- daylight versions @ £0.33m and the blind-fire version @ £0.52m.

ii) Hawk replacement- similar

to the System D in the SAM/fighter working party. @ £2.9m /unit consisting of one surveillance radar, two fire channels etc.

iii) The American SAM D system with anti- aircraft and anti-missile capability @ £35-50m/battery.

A Frequency Modulated Interrupted Continuous Wave (FMICW) radar was assumed, carried in an aircraft like an Andover to provide an Airborne Early Warning system with a warning range of 150-200 nm against low flying aircraft. It was assumed that it would make maximum use of the NADGE facilities on the ground for track evaluation, Command and Control and communications. Cost- £2.25m/aircraft.

Financial Constraints. -In order to contain the study within the bounds of realism, it was essential to make some judgement about the levels of future expenditure on NATO's air defences, we adopted the assumption that the NATO countries would be willing to continue to spend at the general level that had been established in the recent past. It was a notional ten-year investment including the capital costs of the hardware for the system, together with spares and the cost of running it for ten years. On this basis, the 10 year investment for the present 150 Lightning and F104 aircraft, for the 36 NIKE and 44 HAWK batteries all in the 2 ATAF area was assessed at £1000m.

There were costing difficulties. Opportunity costs had to be assessed if aircraft transferred from one role to another. NADGE costs were excluded, as these equipments would continue into the future period. The low level SAM deployed with the ground forces, were included in this budget because it was argued that, although primarily for the defence of the troops, they would contribute to the total effectiveness of the defence complex. (Later in the study, we did include the alternative assumption that the Rapiers with the army would be funded separately.) All of the above emphasised the importance of taking the work of the study group as an indication of trends and possibilities, and giving general guidance on the interplay of the various components of possible air defences.

The Effectiveness of a Range of Possible Air Defences.

The next step in the study was to postulate various air defence mixes within the technical and financial constraints assumed and to display the characteristics of these mixes to assist the appreciation of their effectiveness in dealing with the spectrum of possible attacks available to the enemy within the limits set by the number and characteristics of his Air Order of Battle (AOB). The medium altitude coverage by the SAM systems of 1968 (page112) were broadly speaking, a belt of Hawk systems with an overlapping

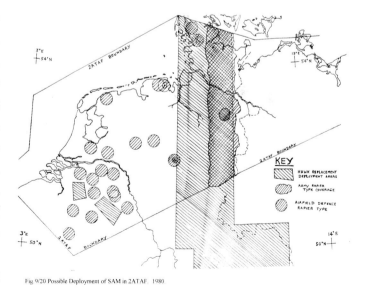

Fig 9/20 Possible Deployment of SAM in 2ATAF. 1980

Possible Deployment of SAM in 2ATAF.1980.

coverage of the Nike systems some 50miles back from the geographical border. There was also some medium altitude cover of valuable sites in the rear. The fighter bases were 150 to 333 km behind the border. In war, the fighters were to provide defence in depth and, in peacetime and periods of tension, policing of the airspace.

The consideration of future equipments and their deployment led us to conclude that the future requirements could be:

i) Hawk replacements, including an anti-aircraft and anti-missile capability of the type recommended in the SAM/Fighter working party and Rapier, would be deployed.

ii) Army Rapier would provide air defence coverage forward with the army and to valuable points,

iii) fighters as described earlier supported by an AEW system deployed as in the diagram below.

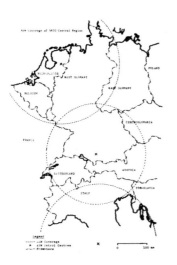

Possible deployment of the AEW Aircraft.

It is very difficult to display the capability of air defences in a simple form, either verbally or pictorially. The range of possible tactics open to the enemy is very wide, the kill potential of the defences is a function of raid concentration, both in space and time, the effectiveness of the defences varies with time in a complicated way with interceptors having to return to base for re-arming and refuelling after each sortie, and the missile batteries having to stand down for brief intervals for reloading and, perhaps for extended intervals, for re-supply after a particularly active period. The overall state of the defences can vary by quite a large factor between the opening stage of a battle, when all systems are at high states of readiness, and later stages when many sorties have had their inevitable effect on aircraft availability and the flow of battle over the ground has lead to missile batteries being out of action, as they move from one location to another, or are knocked out by enemy action. In an attempt to reduce these options to manageable proportions, evaluations were made of defence mixes for 12 typical situations. Three variants of enemy attack tactics exemplifying different choices of concentration in time and space; two depths of penetration into NATO territory and two cases of defence preparedness as illustrating conditions in the early stages of battle and after some days when wastage, disruption, and enemy action have taken their toll.

The volume of information produced was large. However, the most important indications stemming from the analysis can be illustrated in a convenient form. The diagrams refer to a situation in the early stages of operation in which all defences are at a high state of readiness and at full strength. The attack penetrates 100 miles into NATO airspace, along 5 axes across a typical corps front of about 50 miles. The raiders come through in waves of 10 aircraft, totalling 300 aircraft within one hour and attack speed is around M=0.9.

The weaknesses of the 1968 defences are shown in the upper diagram. At attack levels below one thousand feet, or there about, the intruders would get away almost loss free, a very small number of aircraft only falling to gun defences deployed with army formations and to visual interceptions by Lightning and F104 fighters. At altitudes from some thousands of feet to around 50,000 ft, the defences are very strong. Above 50,000 ft, the kill potential falls sharply, with Nike alone having any capability above 60,000 ft.

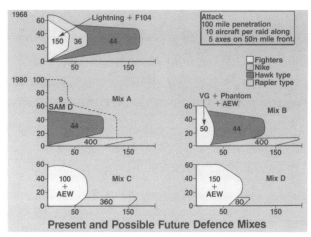

Present and Possible Future Air Defence Mixes.

Mix "A" shows a possible *all-missile* defence involving no fighters. At first sight, the defence mix shows a considerable improvement over the existing system. The weakness at low level has been removed by the introduction of Rapier when there is a high kill potential of over 100 in visual conditions and over 50 in non-visual conditions. Medium altitude defence is good and SAM D might provide a reasonable capability up to 100,000 ft. Yet the all-missile defence has very serious shortcomings. It has no capability for anything other than destruction. The interception and policing functions, which may be of importance in peace and in periods of tension, are beyond it. It is a particularly "brittle" defence in that - once a small number of key missile sites are destroyed by enemy air or ground attack, a gap is created through which the enemy can pass unscathed. If the enemy forces occupy the areas from which these missiles operated, it might be impossible to plug the gap.

In Mix B, the expensive SAM D has been traded for 50 interceptors plus AEW (of the simpler type). Low and medium level defences are still good, but the defence against intruders- probably reconnaissance aircraft -above 60,000ft is now considerably less than in Mix A. In exchange for this cutback in high level kill potential, the system has now a capability, through the interceptor, of rapid reinforcement of areas of particular importance and of plugging any gaps made in the missile belts. The mix also gains a capability for policing and identification in peace and times of tension. There was no doubt that this was a move in the right direction, since the physical threat from the enemy aircraft at 50,000 ft and above was very small. Operational experience, however, suggested that a force of only 50 interceptors within an area of responsibility as large as that of 2 ATAF left very little in hand for the flexibility necessary to guard against tactical concentration and outflanking and possible enemy ploys like " flushing" the defence interceptors, or the inevitable wear and tear of some days of action. Mixes C and D display the consequences of providing, respectively, 100 and 150 interceptors to overcome this shortcoming.

In Mix C, the additional 50 interceptors have been provided at the expense of the medium altitude SAM and a small reduction in the deployment of Rapiers. Again low level defences are very good; there is some cutback in medium altitude defence compared with Mix B; there is a significant increase in the capability for engaging targets- maybe reconnaissance aircraft- at

50,000 ft and above and there is an unquantifiable increase, the full importance of which only experienced judgement can assess, in the general flexibility of the total defence system.

In Mix D, with 150 interceptors, the aircraft component of the system is built up to give still greater flexibility, this time at the expense of the majority of Rapiers deployed forward. The pronounced low-level "tongue" of the preceding mixes is cut back somewhat, but the total low-level kill potential is still quite high; the medium level capability has been increased; and there is some increase in kill potential above 50,000 ft.

As mentioned earlier, it was difficult to present all the factors that must be considered. The study assumes that the NATO interceptors are deployed, for safety, fairly well back and scrambled on receipt of warning from the AEW aircraft. Fighter interceptions near the forward edge of the battle area would be few, particularly if enemy jamming forced the AEW aircraft back when the first indication of enemy attack was obtained as they crossed into NATO airspace. In such a situation, the only protection that the forward units would have against enemy low-level attacks would be that which their own Rapier defences provided. In Mix D, this would amount to a kill potential of around 30 in visual conditions and 10 in non-visual conditions. It should also be remembered that the diagram shows the situation in the early stages of battle, when all systems are prepared and ready for operation. At later stages in the battle the kill potential would be considerably reduced.

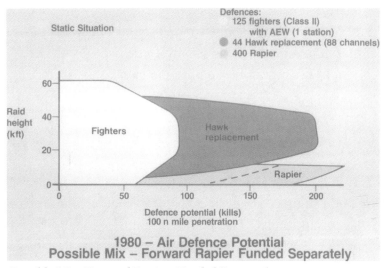

Possible Mix- Forward Rapiers Funded Separately.

In all the mixes, the assumption has been made that the money for all of these systems was found from within the total £1000m budget. However, if the alternative assumption is made that the forward Rapiers are funded from other sources, it would then be possible to have a mix that included 125 interceptors, a Hawk replacement and the independently funded low level SAM. With the extra £400m, the defence is more powerful. In particular, it would give a considerable increase in the kill potential between 10,000 and 40,000ft which could be brought to bear well forward of the army units. This capability would be a valuable counter to enemy tactics of coming in over the Rapier ceiling and then making a brief excursion through the Rapiers to attack forward area targets.

For high, medium or low altitude reconnaissance missions, the enemy could use a small number of FOXBAT aircraft at supersonic speeds and there would be considerable advantage, in terms of reduced losses, for them to do so. The effectiveness of medium SAM falls off against targets flying faster than Mach No.1.0 but should retain some effectiveness up to

Mach No.2.0. Rapier effectiveness is very low against supersonic targets. The interceptors should retain their capability at high altitude against target speeds up to about M 2.4. Against FOXBAT aircraft at altitudes up to 100,000 ft and speeds up to M 3.0 on reconnaissance missions, only the inclusion of SAM "D" would deny the enemy this flexibility.

The final balance between low and medium altitude SAM and interceptors depends on the developing picture of the enemy threat, the costs and capabilities of the possible components of the mix, the extent to which we can get our NATO allies to agree with and accept our ideas and the decisions on hardware and equipment that they make in the years ahead. There can be little doubt that NATO's air defences of the 1980s should contain an interceptor with a good capability at medium and low level, AEW and a good low level SAM system of about the cost and characteristics forecast for Rapier. A replacement for, or an improvement on the medium HAWK system can be neither ruled out nor firmly supported on the basis of this study. The analysis showed clearly the value of the AEW so that the interceptors can be scrambled from their bases in time to make their kills, rather than operate on combat air patrol.

Subsequent Events.

The Hawk and Nike systems in the 2ATAF area in 1968 were manned and deployed by Germany, Holland and Belgium. Any UK systems, Thunderbird and Bloodhound were deployed near valuable targets in the rear. I never discovered how the systems were paid for but it seemed to be that the German, Dutch and Belgium governments acquired them as some sort of offset deal with the USA as recompense for the deployment of US forces in Europe. If it was assumed that this arrangement would continue, the need for a UK system to meet GAST 1210 disappeared. Medium altitude defence of the UK Army in Europe would continue to be provided by other countries. SAM "D" became PATRIOT and was not deployed in Europe.

AEW- The UK regarded the use of Airborne Early Warning Aircraft as a fundamental part of air defence against low flying attackers. In order to persuade our NATO allies that this capability needed to be a fundamental part of the overall European air defence capability, a team from MOD toured Europe. Air Vice-Marshal Steedman ACAS (Policy), Ian Shaw and I visited the main Defence HQs, including in June 1969, the C-in-C Allied Forces Southern Europe to sell the

At H.Q. Allied Forces Southern Europe.

concept. It was not welcomed in Germany where they planned to deploy a multitude of small Siemens radar systems spread all over Germany. We could not see how this system could possibly be operated effectively. AEW was welcomed by Italy where the advantages of it in their air defence environment were clear to them.

Unfortunately when it came to writing the Requirement for the AEW, the air controllers included everything but the kitchen sink so that the aircraft became like a ground radar station with multiple control positions. The HS801 Nimrod system never performed adequately and, when the aircraft caught fire, the development was abandoned in 1986 after 9 years and a cost of £1.1 billion. Europe eventually bought the American AWACS airborne early warning aircraft.

In a gap between studies, I took the opportunity to visit the OA Branches in Hong Kong and Singapore. As I was more or less passing, I took the opportunity to make a four-day visit to the Military Research and Development Centre in Bangkok. Major General Prasart Mokkhaves had arranged for me to be briefed on their work and then on the following three days to be flown to the Mekong River on the border with Laos to look at their problems of tracking and perhaps stopping the Russian helicopter flights into Thailand at night. I found the problems both fascinating and, with the resources that they had available, quite intractable.

Just before the SBAC show in 1968, the International Liaison Office in MOD rang me to ask if I would brief General Westin, Chief of Defence Staff from Sweden and his team on the Air Defence of Europe Study. I agreed and asked for a favour in return. A senior military team from Thailand were coming to UK for the SBAC Show. They wanted to buy some 50 Land Rovers and visit other places. One was RARDE who were refusing to accept the visit. I asked if he could overcome the problem.

The briefing of General Westin went extremely well. The team were clearly fascinated by the study. It resulted in my receiving four invitations to Sweden for further discussions on Swedish Air defence. In March 1969, the UK Defence Sales Organisation sponsored the first visit. The Swedes had arranged a trip to an AA training school on the Baltic Sea to observe their Bofors guns in action. It was an interesting demonstration without many hits on a straight flying towed target. The Baltic was frozen and there was a steady breeze off the sea. I froze despite having all my own warm clothes on beneath the Swedish army winter gear. It required four hot baths before my body temperature felt normal again.

That evening we had a party in the British Embassy where the wife of a Swedish Naval Captain asked me whether I was carrying an arsenic pill just in case I was kidnapped. Such was the conversation at Embassy cocktail parties. The next day was devoted to the briefing and discussions. Over lunch they told me an interesting story of problems in Stockholm harbour. After being old enemies of the Russians for hundreds of years, Sweden now had a neutral, sort of stand off, relationship with them. However, the Russians were sending their smaller submarines on reconnaissance exercises into Stockholm harbour. After they penetrated the harbour, they would sit on the bottom. If they were detected, the Swedes would send down divers to fix a small shaped charge to the outer skin of the submarine. When it was detonated, it punched a small hole through the skins. Whenever this happened, the submarine always left and the Russians never complained. My visits to Sweden were always memorable.

The Maritime Air Protection Task. March 1967.

During the Defence Review, a Maritime Operations Working Party was set up under Deputy Chief Scientific Adviser (Studies) to examine a number of possible shapes and sizes for the future navy. One of those studies was allocated as a task to DOAE under my chairmanship.

Its terms of reference were- to determine the numbers and mode of operation of land-based aircraft which would be required for maritime support protection tasks in the 1970s, on the assumption that there would be no fixed-wing aircraft carriers. The work was separated into three phases: -

i) the operational environment and scale needed to counter the threat of Indonesian attack

ii) the logistic requirements to support the operation, with a very limited examination of the Command, Control and communications and

iii) other scenarios.

The scenario chosen was similar to that chosen by the main working party using a carrier in the task force. It involved the protection of the sea-tail convoy, carrying a Brigade Group for the assistance of Malaysia, while en route from Western Australia via the Cocos Islands and the Malacca Straits to Penang. Subsequently, smaller convoys proceeding via Gan to Penang to re-supply the Brigade group would also require protection. These convoys are assumed to be under threat of attack by Indonesian air and naval forces. The fighters, AEW and tanker protection of these convoys were assumed, as the main task force progressed, to operate from Learmouth, Cocos Islands and Butterworth and, cover the follow-up convoys, from Gan and Butterworth .

At the time of the study, the Indonesian Air Force capability was poor. It had Russian aircraft that were inadequately maintained and showed little evidence of effective operational capability. However, we assumed that if they had aggressive intent towards Malaysia then they would revive the air capability. The Joint Intelligence Committee agreed with us a predicted capability in 1976 and beyond which included Badgers and Brewers, carrying Kipper air-to-surface missiles, and Blinders, Fishbeds and Fitters using direct attack weapons.

The convoy was defended by an Airborne Early Warning Aircraft of the BAC 1-11/HS801 type using FM/ICW radar giving a look down capability against low flying attackers. The fighters were Phantoms using Sparrow air-to-air missiles. The fighters and the AEW were on Continuous Airborne Patrol (CAP) refuelled in the air by tankers on CAP. The SAGW defences carried by the escorting ships were Sea Dart, Sea Slug, PX430 (which became Sea Wolf), Sea Cat and gun systems. The kill rates against aircraft were determined for cases where there were no fighters present and SAGW provided the only defence and for cases in which the fighters achieved some attrition of the attackers before the raiders were engaged by the SAGW defences.

Attacks on the base facilities were also assumed as well as on the supply chain necessary to support the operation.

We had considerable debate on how a task force commander would operate his defences. We decided that he would graduate his defences to counter the estimated threat along the route. This concept allowed us to vary the fighter strength on CAP along the route starting with only two fighters on CAP when near Learmouth, four when operating from the Cocos Islands, and eight over the Malacca Straits. The defence level was set such that in a typical raid an average of one ship would be lost. (This assumption upset a number of our naval colleagues. They argued that they might be on that ship.)

The highlights of the study were: -

i) the AEW is vital to the defence of the convoy and the bases.

ii) Sea Dart is by far the most effective element of the ship defences.

iii) the force requirements of fighters , AEW and tanker aircraft were estimated using a computer model taking account of the endurance, reliability at take-off and in flight, and the time required to rectify the defects and prepare the aircraft for the next sortie. In this way it was possible to estimate the variations in the expected number on CAP with force level and with operating distance from base.

iv) the overall numbers of Air Force personnel involved in the operation, excluding the Transport Command element , lie between 6500 and 7500 men, depending on whether 7 hour or 5 hour sorties are flown and the redeployment plan adopted.

v) the problems of supply have been examined in sufficient detail to estimate the weights and bulk of the spares required.

vi) the overall time required for this operation from the initiation to the arrival of the sea-tail at Penang is 30 to 35 days and 53 days with and without a floating stockpile respectively.

vii) the study at DOAE has confirmed that land-based aircraft could be used in support of maritime operations if a practised organisation achieving the normal standards of efficiency was assumed. The intensity of flying was high and the operations require good communications, command and control. The airfields were crowded, operating 150 UK aircraft of six or more different types from the same base, and therefore need a good airfield organisation and spares supply system. *Regular trials and exercises were essential if this is to be achieved.*

In the wash-up on the study, at the Maritime Operations Working Party, emotions ran high. On the whole, the study came out unscathed but did demonstrate the difficulty of providing air support from land bases. The Navy were determined that they would not loose their carriers. Their proposal was to build a very big carrier, the CVA01, but it could not be afforded within the UK defence budget. The CVA01 was cancelled but a smaller ship, the Through-Deck Cruiser, which looked strangely like an aircraft carrier with a flat top and a hanger deck just below it, was approved.

Exercise HELLTANK.

By this time I had taken on the Field Trials Division under Alan Goode and given up the Land Division.

At the time, the UK lacked an established tactical doctrine for the use of helicopters in the battlefield in the anti-tank or reconnaissance roles. DOAE was asked to conduct a series of trials and exercises to assess the effectiveness of helicopters in European type armoured warfare. Over a two-year period, Alan Goode and his team completed a series of trials and associated studies.

The first three phases, conducted by the Army Air Corps, explored particular aspects of the helicopter's ability in these roles and demonstrated both its potential as an anti-tank weapon and its invulnerability to the main armament of the tank. Phase 1V, from which the bulk of the data was derived, was a 4 day tactical exercise in which an array of modern anti-tank guided weapons and all other elements of armoured formations were realistically deployed over a frontage of 12 km with a depth of manoeuvre of about 60 km. The ground density of the units was arranged to be roughly characteristic of Warsaw Pact and NATO armies. Each force was composed of 1 armoured regiment, 1 mechanised infantry battalion supported by sub-units of the Royal Artillery and Royal Engineers, 6 anti-tank helicopters and 12 reconnaissance helicopters. Each ground force was defended with the following anti-helicopter simulated systems; Eight Rapier systems without radar warning and ranging facilities, 6 Blowpipe, 18 Rarden 30 mm cannon mounted on APC's and 6 Swingfire equipments with appropriate compliments of tank guns, machine guns and rifles. The anti-tank helicopters were assumed to be equipped with manually guided missiles of the SS11 type with a range of 4000metres, but their crews had only stabilised hand held binoculars for target detection and acquisition.

The primary aim was to assess the opportunities for engagements between helicopters and ground forces, and to record the appropriate geometry and duration of each engagement. All ground weapon crews maintained logs of the helicopter sightings, ranges and altitude of flight, their positions at the time and the duration of sightings by stopwatch. The co-operation of the participants and the quality and quantity of the recordings was good.

Over 200 helicopter sorties were flown, resulting in: -

Over 250 attacks on AFVs by anti-tank helicopters,

Over 100 contact reports by reconnaissance helicopters, and

Over 2000 claims were made by ground weapon crews of the engagement of helicopters.

Details of all the attacks and engagements deemed acceptable for technical and tactical realism were combined with the performance data for the relevant weapon. This provided a basis for estimating AFV and helicopter losses had the firing been live.

Caution is always required when translating results of trials involving simulation, however realistically designed, to operational conditions. Nevertheless, with improved avionics and optical sighting systems for the helicopters and improved tactics and control procedures, the exposure and consequent vulnerability of helicopters, it should be possible to keep below the loss levels indicated in this trial. If such improvements were effected the conclusions would be: -

a) Helicopters armed with guided weapons could find and engage AFVs in European-type armoured warfare and be capable of making a major contribution to the anti-armour defence. However, they are potentially very vulnerable to the Rapier type weapon and, therefore, must be flown and committed with prudence. They are likely to prove most effective and least vulnerable when used in highly mobile battle situations.

b) reconnaissance helicopters should be committed with great caution. They are likely to be less vulnerable and more effective when tasked to obtain or confirm specific information.

The Field Trials Division really scored a great success with this trial. It was an excellent example of planning a major exercise with the Army, getting the maximum commitment from the troops, arranging the data recording systems such that the data could be analysed and then used to extrapolate the information to variations of the weapon mixes and tactics. Finally it is essential to issue the report and make numerous presentations so that valuable lessons could be introduced into our operational thinking.

Visitors to the Establishment

We welcomed many important visitors to the Establishment. One such visitor was Prince George of Denmark accompanied by Lt. General Blixenkrone-Moller, Commander in Chief of the Danish Army. Prince George was briefed on the overall objectives and work of the Establishment and his C-in C on the war-gaming activities.

(Photograph by kind permission of the Surrey Herald Newspapers.)

<u>House of Commons Select Committee on Science and Technology</u>
<u>- Defence Research. Session 1967-68.</u>   2nd May 1968.

By this time, I was temporarily Director of DOAE as Terry Price had moved on to an atomic energy appointment.

Sir Clifford Cornford CDP, Brigadier Lewis Deputy Director Military at DOAE and I were called as witnesses before the committee. We were invited to provide an introductory statement that gave information on the matters to be investigated by the committee. Some of the committee were friendly but probing, others were distinctly hostile. Nevertheless, we seemed to give a good account of ourselves and everyone understood a great deal more about what we did as a result of the question and answer session.

The next day, the Daily Telegraph wrote a short piece- DAYLIGHT on DEFENCE. The first paragraph read: - "It is very hard for those who have had operating responsibilities in Government to view without bias the selection of alternatives which may call into question their own prior accomplishments and choices." This phrase comes from a letter written by the Rand Corporation of the United States to the Parliamentary Select Committee on Science and Technology. It explains why less than 3% of the staff employed on the Corporation's "War Games" studies were serving officers. In comparison Britain's DOAE at West Byfleet employed some 30% of serving officers. Byfleet was so wrapped in secrecy that the Select Committee learned relatively little more about it. But the committee members have heard enough about the various defence establishments to feel, some of them strongly, that the more defence policy is exposed to outside evaluation the better. Whilst its report to Parliament later this session is unlikely to be dramatic, Defence Ministry witnesses have discovered that it can and does demand detailed written answers of a sophistication far beyond normal Parliamentary questions. ———— The standard of defence debates are notoriously low. At the very least the 14 MPs on the Committee will now be able to improve that standard."

At the end of the session, the Chairman suggested that he had gathered that we did not have much influence on defence matters!!! Sir Clifford Cornford replied that many of the problems that were being studied bear absolutely directly upon the proposals that the Weapons Development Committee has to consider. Inevitably, they are in the highest degree speculative when one is forecasting ten years ahead. —- What one is tending to look for is a robust solution that would be least affected by changes in the assumptions that one has had to make. *We are not court wizards!*

On 15th October that year, Sub- Committee A of the same Select Committee visited West Byfleet. I, with Andy Stratton the new Director and others, appeared before them for some two hours. This session dealt, in much more detail, with issues of staffing, how we were tasked, what influence we had on the specification of those tasks, whether we should be outside government as Rand was, how we worked with other establishments and industry. I felt that it was a very successful visit for us.

Farewell to DOAE.

Having welcomed in the new Director, it soon became clear that he was going to take a new approach to the work. He seemed to me to want to take time off from doing studies and start building new computer models which would be designed to answer any question put to us by MOD. In addition, the Establishment would be reorganised on the basis of two main divisions, Land-Air and Sea-Air, to match the way military engagements take place. We would be doing all multi-service studies and no longer undertake the single service work that, although only a small percentage of the effort employed, helped greatly with our relationships with the single service Departments. In order to provide the necessary effort to create the computer models, we should wind down our current study commitments. I felt that the policy was wrong, but most of all, I could not default on the commitments that I had already made to both the Operational Analysis Committee and to the single services.

After he had settled in for about a month and visited all the Divisions and been able to discuss their work, he come into my office one morning to discuss the new arrangements. It was clear that he had not been briefed on some of the single service work that we were doing. As I explained it to him, he gradually became paler with a deepening frown. Ultimately he got up, told me that I was in the wrong job and left the office.

After some difficult times, it was decided that I should go back to RAE to Head the Weapons Research Group in Weapons Department. At that point, Stratton decided that he would retain me at DOAE until the personnel system had found a replacement acceptable to him. In the end, Andy Smart from RRE was chosen. Unfortunately the impasse lasted some three months. I was greatly relieved when it was brought to an end.

Chapter 7

Head of Weapons Research, Weapons Department. RAE.

After many weeks trying to run areas of two Establishments part-time and keep my head below the battlements, it was a relief to settle in at Farnborough. It was an even greater pleasure to receive a letter from Morien Morgan welcoming me back.

I looked forward to getting back to the technology and project issues. The technology was developing fast and there was much work to be done to keep the UK weapon technology in the forefront. At the time, we were just starting the early experiments on helmet sights for weapon aiming and decided that Avionics Department would be best able to undertake this programme, as it would need to be heavily integrated with the overall cockpit technology and pilot-machine ergonomics. Low light TV was being overtaken by infrared technology. RRE had researched and developed, with industry, the sensors/detectors for infrared and were concentrating on its development for scanning aerial reconnaissance.

At Farnborough, we wanted to develop a matrix of I.R. cells so that when installed looking forward from an aircraft or missile a complete picture of the terrain and targets ahead could be shown without scanning. When I arrived Mike Weedon already had some detectors working using a small number of cells in a matrix. To make it of military value, the number of cells had to be greatly increased to improve the quality of the picture and the output fed to a TV raster that would be compatible with the presentation of other information in the cockpit. In such an active Department, there were many concepts that were being explored. I will concentrate on the anti-airfield weapon and give a complete history of the programme. I was involved with it from time to time until it was taken out of service in the mid 1990's.

JP233

The Air Staff felt that they needed a weapon to reduce the effectiveness of the Warsaw Pact air capability in Central Europe and were working with Weapons Department to examine the options available. They had written a draft Staff Target (Air) 1217 to prevent take-off from runways and other prepared take-off areas. I remembered the problems generated by the lack of a complete operational cost/effectiveness study on an earlier weapon and I wanted to ensure that this part of the work was given a high priority very early in the feasibility stages.

In WW2, Herman Goering launched the Eagle Offensive. He intended to destroy Fighter Command in 4 days. There was a continuous offensive against airfields and aircraft on the ground in southern England. Airfields like Biggin Hill could be unusable because of craters at lunchtime but in use again by late afternoon because of the speed with which the grass runways could be repaired. In the Arab-Israeli war of 1967, surprise attacks were made on 19 Egyptian Airfields. The runways were bombed to stop aircraft from taking off. The aircraft were then attacked on the ground to destroy them, thereby achieving air superiority. The result was a relatively easy victory by the Israeli army.

The lessons of this war were soon learned. Shelters appeared on Soviet, Warsaw Pact and NATO airfields. Shelters are very tough targets and it is very difficult to know whether they are occupied. The W.P. countries were known to have capable on-base repair facilities and their practice was to repair the runways by overlaying an additional layer of concrete. The Warsaw Pact airfields were, therefore, very tough targets. The issue was how to attack them so that aircraft using short take-off techniques, like the Harrier, could be prevented from taking off.

For attacking the surfaces three options were available.

i) use small shaped charges like BL755 for chipping the surface. Unfortunately, the surface can be swept of debris and repaired by using a skimming of quick setting concrete.

ii) try to produce a crater of at least one metre diameter and, if possible, uplifting the surrounding concrete to make the repair difficult by ensuring that the concrete slabs around the crater have to be removed to make a satisfactory repair.

iii) use a large weapon of say 1000lb to produce a four to five metre crater. This is very effective but the number of craters that can be produced by a given number of sorties is small. (This technique was used in the Desert Storm using Paveway laser guided weapons on runway intersections to achieve the greatest disruption per weapon used.)

Our view was that the one-meter crater was probably the best compromise.

SG357 descending on a parachute.

Work was undertaken by RARDE at Fort Halstead, AWRE at Aldermaston and overseen by the Weapons Project area. This combination was a very powerful grouping containing the best warhead designers in the UK. The design of the runway cratering weapon, designated SG 357, left, descending on its parachute had a large primary shaped charge warhead situated at the rear of the weapon firing a plasma jet through the centre of the secondary warhead-upper part of diagram. The plasma jet creates a narrow hole in the runway of one meter or so deep and as the hilt of the plasma jet "sword", travelling at about 9000 ft/second, gets to the secondary warhead, it pushes it into the hole.

The secondary charge has to be pushed past the out-rush of debris from the hole, avoid being bounced out of it and survive intact with its arming and fusing mechanism still effective-lower part of diagram.

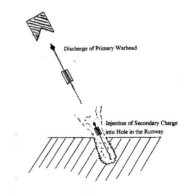

The Detonation of the Primary Warhead and the Injection of the Secondary Warhead under the Runway. SG357

The SG357 Crater, Uplift of Concrete Slabs and Debris.

There it lies for a few seconds before exploding to produce the crater, the debris and uplifting the concrete slabs around the crater. The programme had reached the early experimental work when I left it to a post in London (described in the next chapter) but, as I was involved in the programme several times later, I feel that it would be interesting to describe it through its development, in-service maintenance, its operational use in the Gulf War Desert Storm and the reasons for it being taken out of service in the late 1990s.

In November 1987, I lectured to the Royal Aeronautical Society on JP233. In it the whole weapon was described including HB 876 area denial munitions (right). They sit erect on the surface, were obvious and were intended to delay the airfield repair teams from doing their job. The two main elements of the weapon, shown on the right of the picture, were ejected from the dispenser.

The Two Components of the Weapon.SG357 & HB 876

JP 233 Ground Coverage of an Eight Aircraft Attack of an Airfield.

In the operational concept proposed for the European Theatre, a raid of eight aircraft could produce 480 craters and sow 3440 area denial mines to give a ground cover in which the lines of craters are surrounded by a field of mines. At the time, there was not a more effective weapon in the world.

The mines are a very important element in the effectiveness of the weapon as it is vitally important to delay the work of the repair teams. What could not be said at that lecture, related mainly to the intelligence built into the mine. On landing, it jettisons the parachute, erects itself and switches on. The fusing logic of the mine contained anti-disturbance and event timing programmes i.e. it switches itself on and off at random times as a countermeasure against it being detonated by disturbance caused perhaps by shooting at it from a safe distance and, at another random time, the mine explodes within its total life of 6 hours. It seemed desirable that its life should be longer, ~24 hours, but the battery life dictated the maximum that could be achieved and still be confident that all the mines would self-detonate.

The Gulf War.

First task of the allies was to achieve air superiority by: -

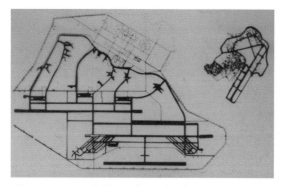

Comparison of Typical Airfields, Warsaw Pact (upper right) and Iraqi.

i)   taking out medium and high level SAM defences

ii) destroying as many aircraft as possible in the air,

iii) closing the airfields and then

iv) destroying the aircraft in shelters

JP233 was designed for task (iii). Unfortunately the Iraqi airfields were 10 times the area of the Warsaw pact airfields. Strike Command was asked by the Command in the Gulf to suggest optimum methods of using weapon. At the time, I was a Director at Hunting Engineering and had several discussions with the Command Operational Research Officer on the options that might achieve the best results. With the very limited resources available, (4 aircraft per airfield) in Saudi Arabia compared with eight for much smaller Warsaw Pact airfields, it was recommended that the weapon be used to isolate the shelters from the runways and taxi tracks so that the aircraft could not be operated or escape.

Subsequent to the attacks, all reconnaissance evidence indicated that none of the Iraqi aircraft moved from the shelters isolated in this way and none of the Tornados were lost on JP233 missions. Ground inspection after the ceasefire showed that the weapon had worked as planned and all the mines had exploded.

As these are "one shot" weapons that might sit in storage for fifteen years or more, to be sure that they would work if ever they were required, in-service proving trials were held every year at the West Freugh Trials Range. At random, one weapon was selected for a strip examination and another was selected for dropping over the range. They were successful and gave confidence that, when required, they would work.

In 1997, the Red Cross started to campaign against anti-personnel mines and, with the very high profile support of Princess Diana, it became a very emotive media circus. In these campaigns the truth or the full story very seldom gets any hearing. There is no doubt that the casualties in Angola were horrific but still tiny compared to those attributable to the AK47 or the machete. What was never said was that the United Nations had been working on a study to identify those mines that should be banned and those that were needed for military operations. They recommended that an acceptable mine for military operations should automatically deactivate and meet the criteria that: -

-in one month 70% should deactivate

-in three months 99% should be inert.

HB876 meets these criteria. In fact, as described earlier, the mines sit on the surface, are large and visible and are programmed to self-destruct within six hours. They are sophisticated and as a result are expensive.

Contrast that with the mines sown in Angola, and most civil wars. (In Yugoslavia, they played a major part in preventing successful Serbian artillery attacks on Saryavo.) The mines are very cheap to produce in small workshops from components that are readily available in the world markets. One of the important ones is the pietzo-electric activator that operates by producing a significant voltage and current when a small pressure is applied. It can stay effective indefinitely. The pietzo-electric element is used by magicians to light neon bulbs on peoples heads and even by shoe manufacturers to produce flashing lights on children's shoes. The main production of these mines is in China, Russia and Pakistan, something like 1/3 of the world's population. Aid from the West into civil war areas is often used to purchase these weapons. The countries that produce them were not prepared to sign a ban. The USA was prepared to support a ban on mines that did not meet the U.N. criteria. However, they insisted that the high-technology mines were a fundamental part of its commitment to secure the exclusion zone between North and South Korea.

The campaigners would settle for nothing less than a total ban.

In May 1997, the UK decided to support the ban and destroy its stock of anti-personnel mines by 2005. The ban would also apply to JP233. Many army officers said that it was absurd to equate the responsibly spaced and carefully mapped mine laying of the Western armies to the randomly laid minefields that, every year, kill and maim many civilians in the third world.

The UN estimated that there were 15 million mines in Angola, the Red Cross half that number but the estimates by the de-mining agencies was less than 500,000, a factor of fifteen below the estimates of the Red Cross. The mine clearing groups say that the numbers had been exaggerated to such an extent that it has deterred the public from making donations, thinking that the problem was intractable.

The Red Cross policy itself was inconsistant as the ban allowed: -

a) the use of Schielder vehicle scatter-launched anti-tank mines that were sensitive enough to be set off by stepping on them and
b) the use of Claymore anti-personnel mines because they were not victim actuated but by soldiers when they wish to attack a target. It was, nevertheless, very easy to modify them.

It is doubtful whether adequate consideration was given to the needs of our own troops. Consider an example of being on a patrol in the jungle, when some anti-personnel mines laid around the camp might enable the troops to get some sleep without fear of surprise attack. There are many other uses where anti-personnel mines contribute to the protection of our own troops. This useless political gesture will just cost us lives without stopping the manufacture or use of a single simple civil war anti-personnel mine.

<u>West Freugh</u>

I was working quietly in my office one day when the Director, Sir Morian Morgan rang me and said that he wanted me to go to a meeting the next day in the Cabinet Office and *save West Freugh!* My first question was "From whom"? The Army had been studying all possible sites around the UK for an artillery ammunition testing range for shell proofing.

They had a team of three officers on it for two years and had decided that West Freugh was the most suitable site. Because it was an intra-government dispute between the Army Department and the Ministry of Technology, then the arbiter would be the Cabinet Office.

The West Freugh range is based at the northern end of Luce Bay, some 8 miles south of Stranraer and covers a sea area of 140 square miles together with a land area of 5 square miles, an airfield with stores facilities, target and other land range facilities. Marine surface targets are also available. The range is primarily but not exclusively devoted to trials with aircraft and airborne weapon systems, for which complete facilities, including instrumentation were provided. More than 1000 trials were mounted annually including, research, development, evaluation/acceptance, Service practice with bombs and other stores, parachutes and flight systems as well as navigation training for Service crews. The responsibility for the site belonged to Instrument and Ranges Department with Bill Pye, the very competent Head of Department. Unfortunately, he was wheelchair-bound and therefore very limited in what he could do with respect to meetings in London. He, plus Dai Rees, the Officer responsible for that range, were asked to lend me all possible support. This effort they gave in full measure.

I was soon to discover that technical understanding was not a high priority in the Cabinet Office as the Army had already convinced them that firing 1000 shells in one week was equivalent to one year of air trials. Therefore, their proposals for the use of the range must be 50 times more efficient than its current use for air weapons. It was two weeks of uphill struggle to convince everyone that such a simplistic approach was invalid. Eventually I succeeded and everybody accepted that this unique air range was essential to the air capability in the UK. The Army had to settle for one of their other choices. I didn't stay to see which it would be. All of the RAE staff involved had a celebratory dinner to mark the occasion.

One Thursday afternoon, Norman Coles, who was then Deputy Controller of Establishments (C) managing the resources and programmes of the Royal Aircraft Establishment, the National Gas Turbine Establishment and the Aeroplane, Armament and Equipment Establishment etc., rang me and asked me to go to London and see him the next day. As we lived only four houses apart, I felt that there might well have been an easier way to do it. However, it was a very formal interview at which he said that he wished me to go to London to be his deputy dealing with the rationalisation of the Research Programmes and of the Research Establishments. It would involve promotion and I had until Monday to give him an answer. Sad but the opportunities and promotion were too few. I had to go.

<u>Chapter 8</u>

<u>Research Establishment Rationalisation</u>

The prospect of the job in London did not fill me with great joy. The first three days on the train convinced me that I had to travel first class. Then, I could work or sleep without being squeezed and still have a high probability of getting a seat. As always, I received a letter from Sir Morien Morgan that was very encouraging. His suggestion that I did not staying in it too long was well received but I had no idea how to implement it. However, having a task that had to involve moves of work and some staff losses, it was nice to know that I would have a reasonable hearing in the RAE where many of my problems would be located.

Sir George Macfarlane was Controller of Establishments and Research, (CER). As the title suggests, he was responsible for the Establishments and the Research Programmes as well as the Project Support activities of the Establishments.

CER had three deputies to assist him. Their responsibilities were: -

Basil Lythal, DCER (A) - responsible for: -

Major Fields of Research	Establishments
Ships and Submarines	Admiralty Experimental Works
Hovercraft and Hydrofoils	Admiralty Research Laboratory
Undersea Warfare	Admiralty Surface Weapons Estab.
General Electronics	Admiralty Surface Weapons Estab.
Radar and Surveillance	Naval Constructors Research Estab.
Communications	Royal Radar Estab.
	Signals Res. and DevelopmentEstab.
	Service Electronics Res. Lab.

Bill Penley DCER (B) - responsible for: -

Air Flight Guided Weapons	Chemical Defence Estab.
Rocket Motors and Explosives	Explosives Research and Dev. Estab.
Conventional Armament	Microbiology Research Estab.
Armoured Fighting Vehicles	Mechanical Vehicles Exp. Estab.
Army Engineering Equipment	Royal Armament Res. & Dev. Estab.
Biological Warfare	Rocket Propulsion Estab.
Chemical Warfare	
Atomic Weapons Res. Estab.	

Norman Coles DCER (C) - responsible for: -

Aerodynamics and Structures	Aeroplane & Armament Exp.Estab.
Aero Engines	Admiralty Materials Laboratory
Navigation and Avionics	Aircraft Torpedo Development Unit
Space	Central Dockyard Laboratory
Materials	Nat. Gas Turbine Establishment
Human Factors	Royal Aircraft Establishment

Other laboratories controlled by the Service Departments and Director General of Ships and the Quality Assurance Laboratories were also included for consideration in the amalgamation proposals.

Each Deputy Controller had a Director General, Establishments, Research and Programmes. David Parkinson, Bill Lord and myself each respectively responsible to the Deputy Controllers for specific rationalisation studies and for chairing the quarterly progress meetings for the research, development and project support activities within the Establishments.

The aim of the Defence Research Programme was to provide the base for development projects, both national and collaborative. It was carried out in the Government Establishments, industry and the universities. At that time the funds available for defence as a whole were limited to 5 1/2 % of the GDP. About one third was spent on R & D and about one sixth of that on research. The programmes had to be agreed by the customers and related specifically to the important Military Objectives set by the government. The ratios of intra-mural to extra-mural research varied widely from 99 to 1 for Chemical and Biological Defence to 1 to 3 for Gas Turbine Engines.

Generally, there was a greater demand than the resources, in manpower and funding, would allow. Every year, proposals for the programmes and the funding were submitted. In addition, a study was required showing the change that might be generated by a 20% increase or decrease in the resources allocated. It was essential to avoid a situation where the resources were too thinly spread resulting in a high probability that all programmes would be too late. Enough resources had to be deployed on the most vital programmes to ensure that they had a high probability of timely success. As a result, some programmes had to be dropped. It was the only way to preserve those programmes with the highest pay-off or having the greatest priority.

The proposed programmes were set out by the Research Establishments, in a format that showed - the objectives of the total package with estimates of the deployed intra-mural manpower over the past year, the current year and the next two, with the predicted intra-mural and extra-mural spend. The Package was then broken down into Work Items. There was always something of a conflict between, on the one hand, the scientist writing the proposals, who attempted to write them in the most optimistic way, speculating on a significant advance in the technology to further the military objectives and, on the other, the Research Manager, who wanted to see specific milestones with dates so that progress could be measured. Inevitably, at the leading edge of science, many proposals do not achieve their predicted potential. Then the earlier the work is stopped, to allow other proposals to be advanced, the better. On the other hand, the programme might be achieved with more or different resources. These issues were discussed during the programme reviews and subsequent follow-up meetings. To get the balance right was never easy.

The Research Programmes were reviewed every year at a series of meetings from April through June, chaired by the DGERPs as appropriate. Local Departmental meetings were held at more frequent intervals. The overall Establishment Reviews were held in July and August each year, again chaired by the DGERPs as appropriate.

Quite separately, other organisational battles were in progress in MoD. The pressure was on to achieve a position where the Controller posts for Sea, Land and Air Systems were all to be filled by Service Officers. The Department of the Controller of Guided Weapons and Electronics was the exception. It was to be broken up and distributed between the other Controllers. In addition, there was disquiet in MoD on the escalating costs of the Concorde project, eventually leading to the formation of the Procurement Executive in MoD and handing over the funding and control of the Concorde to Department of Industry. This split added to the management complications for me because quite a proportion of the aero and space research applied to both civil and military activities.

Research Establishment Rationalisation

At this time, there were many who, in a period of the thawing of the cold war, believed that too much of the countries resources was being devoted to defence.

The new task that CER had been set was to draw up plans for consideration by the Procurement Executive Board that would lead towards a new deployment of R & D resources that matched, in the most economical and flexible way possible, the likely future needs of the Services for weapons and equipment. There were no targets set nor were any limitations placed on the possible changes to be considered except that significant savings were required. The Atomic Weapons Establishment and the Chemical Defence Establishment were sufficiently unique to be left outside this remit.

I was sent on a two week Seminar at the Civil Service College. We were addressed by a wide variety of very senior politicians and civil servants. During that period, I somehow managed to upset both. I can remember Barbara Castle, something of a fire-eater at the time, saying, "The greatest danger to democracy was an efficient civil service that was highly aware of its own excellence. It was the best spying organisation that anyone had known. The Minister could be cut off from the sources of information." George Brown gave a highly political presentation on all the great deeds done by his government. When I asked him why so many seemed to achieve the opposite result to that intended, he was not well pleased. Sir Keith Joseph was by far the most stimulating. The task of the civil servant was to advise on achieving the political objectives and the Ministers job to understand human nature. He did well in presenting both sides of the arguments. A senior civil servant, Sir Anthony Part, felt that Ministers were in departments for two to three years only and touched them only tangentially. He felt that the weakness of the civil service was lack of Ministerial control and excessive autonomy. It often confused rules with tasks. Many civil servants regarded Ministers as birds of passage-here today and gone tomorrow. Another senior civil servant castigated us for inadequate financial control. I pointed out to him that until there were proper delegated powers on finance and manpower, there was no hope of achieving better control. He was not amused. However, by the end of the course, I did have a better understanding of the relationships and the pitfalls in the higher levels of administration.

In addition to my normal DG responsibilities, the Directorate of Trials and Guided Weapons, DTGW, Air Cdr "Fitz" Fitzpatrick, reported directly to me. That Directorate looked after the provision of targets and drones for the air ranges in the U.K. and for the Joint Project with the Australians at Woomera. This task included arranging a monthly freight/passenger flight by a Britannia of Monarch Airlines to and from Australia.

The Airfields Study.

My first rationalisation task was "to examine the requirements for and the utilisation of the four largest Procurement Executive Airfields with the view to reducing them to three or two and make recommendations." The four were Farnborough, Bedford, Boscombe Down and Pershore. There were other airfields attached to ranges but these were left for consideration with the future of the ranges. The four airfields were different in themselves and in their use.

Farnborough had served as a research and development airfield since before World War 1. It was an integral part of the work of RAE as well as having lodger units using it including the Institute of Aviation Medicine, the CAA Air Accident Investigation Branch, an Army helicopter squadron and the Meteorology Research Flight. Although the research departments were at Farnborough and it was very convenient to have the aircraft close to the scientists, most of the experimental work did not depend upon the airfield itself.

Bedford was an "A" Class airfield with a long wide runway, invaluable for experiments specifically using the airfield. This work included accurate performance measurements and the safe operation of difficult experimental aircraft for Aero Flight. In addition, the Blind Landing Experimental Unit, the Naval Air Department including its Experimental Steam Catapult and arrester gear were located there. Bomber Command had four Quick Reaction Dispersal Pads plus buildings near the east end of the runway as part of its Quick Reaction Alert dispersal system.

The Aircraft, Armament and Equipment Establishment at Boscombe Down was responsible for testing all military aircraft and airborne systems, to assess them for their suitability for the Armed Services and was close to most of the range facilities required for that task. The runway was large but had something of a hill in the middle of it, making it unsuitable for precise performance work. It also housed the Empire Test Pilots School and their aircraft.

Pershore was a small airfield, about 8 miles away from the Royal Radar Establishment, and used for housing the aircraft involved in researching and developing aircraft radar, infra-red sensors and other aircraft or helicopter carried equipment.

I asked each Establishment for the utilisation data and the case for retaining the airfield. Generally, I received the sort of replies that I expected, except from Pershore where the number of movements recorded by air traffic control was many times higher than expected. They had some eight Canberras for the experimental work and each of them would spend a great deal of their time on the ground having equipment fitted. To achieve many movements would be difficult. Some investigation revealed that the air traffic control statistics included practice approaches under radar control for a local flying club. They were being done, without charge, to keep the Air Traffic Controllers in practice.

From an examination of the programmes, the costs and likely penalties of closing and moving the work, Pershore became the most obvious candidate for closure. The next question to be addressed was where should the aircraft be located?

Because of the synergy of the work with the work at Farnborough, a move to that location seemed the most obvious choice. However, it was the last place that RRE wished it to go. In the beginning the arguments against Farnborough were as spurious as the Air Traffic statistics. They included arguments such as- Surrey and Hampshire were heavily wooded and the movement of the trees in the wind would interfere with the radar experiments- there would be interference from other electromagnetic sources and -it was further to travel than to Boscombe Down. These arguments were readily countered. Eventually, they found that the Concorde Structural Test Facility depended upon a computer-controlled system for applying temperature, both high and low, and structural loads that represented those applied by the pilot and by atmospheric turbulence. Everyone then argued that, as the proposal was to put aircraft using new high powered radars on the airfield, there was a serious risk that electro-magnetic pulses could get into the wiring and circuitry for the computer control and the Concorde structural test specimen, (which was almost a complete aircraft), could be seriously damaged. Structures Department at RAE quickly took up the same cry.

At the time, the facility programme was in the midst of a long test run to get the simulated hours flown well ahead of the aircraft flying hours. A compressed flight cycle was being used, and there was no possibility, in the near term, of doing any tests to evaluate the risks that the new radars might bring. Therefore, Farnborough, as a new location for the Pershore work, had to be dropped. (Much later, when the test facility was stopped for inspection of the structure, I managed to get one of the high-powered radars near and actually inside the test facility building. No extraneous pulses were ever picked up). In my own view, Boscombe Down was already a busy airfield and there would be times when the NATO war role allocated to it would stretch the airfield capacity to its limits. On these grounds, it was not a suitable location.

In order to try to keep the experimental aircraft near to Malvern, I put a paper to the Air Council seeking their agreement to put RRE aircraft as a lodger unit on an operational airfield such as Fairford. Unfortunately, this option was turned down. Eventually, it was agreed that the trials aircraft at Pershore would be located at Bedford. Because the Property Services Policy, at the time, had a policy of erecting buildings only of the highest quality and brick built, the new buildings erected were of a much higher standard than any other office and laboratory on the Bedford airfield.

The Procurement Executive Air Fleet.

The P.E. Air Fleet was quite large. It included: -
i) research aircraft used by both the government (Farnborough etc.) and industry
ii) development batch aircraft used in the process of the clearance to service
iii) aircraft of the Empire Test Pilots School,
iv) transport aircraft , e.g. the four Devon passenger aircraft used by Farnborough for transport between the widely dispersed parts of the establishments, and
v) the target drones used for weapon proving in the UK and in Australia.

It was not un-natural for the RAF to ask in times of financial stringency, "Why does the P.E. have so many aircraft?" With the considerable assistance of the Establishments whilst under real pressure, we were able, by combining programmes on the same aircraft, to trim a few ex-bombers off the list. During both of the above studies, I had lengthy and valuable discussions with Sir Morien Morgan. We always found a sensible solution.

The Aerodynamics Department of the National Physics Laboratory Teddington.

NPL had been a significant contributor to the science of aerodynamics for many years and the theoretical and wind tunnel staffs were very highly regarded. However, the bulk of the work of the Establishment was on National Standards and new technology was making large demands in this area. There was also a need to upgrade their aerodynamic facilities. It was decided that the optimum solution would be to combine their effort with that of RAE and implement some rationalisation in the process. The bulk of the theoretical staff were transferred to Farnborough and integrated into the appropriate Divisions of Aero Department and the wind tunnel staff transferred to Bedford and integrated appropriately there. Analysis of the future facility requirements indicated that only one wind tunnel needed to be moved to Bedford. The remainder of the NPL tunnels were made available to universities and technical colleges. The whole transition went smoothly without any real difficulties.

Naval Engineering

The Navy had a large number of small laboratories scattered over the south of England. It seemed highly desirable to reduce the overheads in these labs and that they be combined with other appropriate establishments. Those of most interest to me were: -

i) The Admiralty Engineering Laboratory at West Drayton was engaged on the development and evaluation of a wide range of mechanical and electrical equipment and systems fitted into RN ships. It employed specialists in mechanical, electrical and electronic engineering, physics, computer and control science, chemistry and metallurgy. The principal facilities provided for analogue and digital simulation of ships' propulsion and electrical power distribution, development and evaluation of diesel engines up to 4000hp, shock and vibration testing of equipments up to 2 tonnes, magnetic ranging and AC and DC testing up to 17.5 MVA and 120kA capacity respectively. Some of the installations were massive. For example, the two tonne testing-machine was a major installation that could not easily be moved and was the largest shock machine (delivers 70g) available in the UK. It was essential for the demonstration that the shock mountings supporting the equipment were effective and that the ship or submarine would not loose mobility or fighting efficiency as a result of a near miss underwater explosion.

ii) The Admiralty Marine Engineering Establishment, Haslar was a direct advisory service to Ship department at Bath on a broad range of marine engineering. The work covered the performance evaluation, endurance running of auxiliary machinery and plant for warships, gas-turbine fuel system component work, desalination, fire hazard, waste disposal etc.

iii) The Admiralty Oil Laboratory, Chobham handled fuel, lubricants and hydraulic fluids etc. for the R.N. To do that task they had a range of equipment including diesel engines, mechanical rigs and analytical equipment including scanning electron microscopes.

N.G.T.E had a considerable engineering capability and included in its tasks was a facility for endurance testing of ships gas turbines and the testing of their air intake seawater filtration systems. The Navy was therefore familiar with and happy to work with NGTE. With this background, it was accepted by the Navy that a greater efficiency of operation could be achieved by integrating AEL, AMEE and AOL into the management of NGTE and move as much as was efficient and economic to the Pyestock site.

<u>The Director of Trials and Guided Weapons and the Joint Project in Australia</u>

The DTGW branch was headed by a very effective Air Commodore "Fitz" Fitzpatrick and supported by a small staff. It was responsible for the provision of the Guided Weapon Test Range Equipment that was operated by RAE and in the provision of targets against which the guided weapons were fired. This included missile targets, Jindervick radio-controlled targets (at £100,000 each) and the piloted and radio controlled Meteor drone targets (at much greater cost). They were operated from Llambedr by contract staff controlled by RAE from the Aberporth Range. DGTW was also responsible for the monthly freight and passenger flight of a Britannia to and from Woomera, Australia for the trials on the range.

The provision of targets in the most effective and economic fashion was always a problem. Almost all of the initial surface-to-air missile trials were done using Jindivivk targets with miss-distance indicators installed and missiles without warheads but with instrumentation in its place. Mathematical models or the whole guidance system were then generated that matched the performance of the missile system. The extremes of the flight paths and the missile/target end games were then simulated (usually by Surface to Air Weapons Projects Division at Farnborough and sometimes by the companies) saving a considerable amount of money and time. Sometimes the trial missiles made direct hits resulting in the loss of the target. In the end, the whole engagement had to be fully demonstrated. If the project was successful, the trial always resulted in the loss of the target. DTGW was responsible for ensuring the supply of these targets including the "droning" of ex-service aircraft. They had carefully stored a number of DH 110's and these needed to be fitted with the necessary droning equipment.

The Woomera Range had played a leading role in the clearance of surface to air missiles such as Thunderbird, Bloodhound 1& 2, Sea Slug and Sea Dart, in the clearance of air to ground weapons, including Blue Steel, and in upper atmosphere and re-entry research. At that time the trials on the high altitude research vehicle Falstaff, as a forerunner to Chevaline, were essential. However, it was already clear that, with the conclusion of the high altitude work, the need for the range was disappearing. It was not until 1980 that it was closed.

At that time, separate studies were underway on the break-up of CGWL and its reallocation to the other Controllerates. I was called to a meeting at which I described the activities of DTGW. I was asked for recommendations on the reallocation of the resources. I advised them that I could see no way to make a sensible split and that I recommended that the whole should be put into one area. It had the most surprising effect on the meeting. Nobody wanted it. They recommended that it stay where it was, unchanged.

Other Tasks

In the electronics area, under Basil Lythall, advantages could be seen in bringing together the Army Signals Research and Development Establishment (SRDE) located at Christchurch and the Royal Radar Establishment, (RRE) located at Malvern. The integration of the programmes would have long-term scientific and staffing advantages as well as cost reduction benefits. It was not unnatural that the staff at Christchurch saw all the family and environmental disadvantages of the move. Some of the negotiating meetings with the representatives of these establishments, who did not want to move at all, went on very late into the evening. At one particular meeting, with the three of us supporting CER, we were being accused of unnecessary moves, arbitrary sackings and fictitious cost savings. At that moment, my alarm wristwatch went off. (It was set to wake me up before the train got to Ascot so that I changed trains.). We all ducked under the table because we thought that it was a bomb. Ultimately the proposal went through and RRE changed its name to the Royal Signals and Radar Establishment, RSRE. In the final outcome the buildings and hangers, vacated by the move of the aircraft to Bedford, were used extensively for laboratories etc. in the move of SRDE to RRE.

Norman Coles responsibilities for materials research led to an investigation of the merits of bringing all materials research under one management. Although the benefits could be seen, the required changes of responsibilities and objections from the Service Departments were too great to justify the reorganisation at that time. (Some 28 years later, these arrangements have been put in place)

In the end, we had proposals to reduce the number of Research Establishments from twenty four to twelve. If accepted, they would achieve some significant cost saving and, at the same time, the research and development programmes would remain substantially intact.

In the push to improve the efficiency of operation of MOD as a whole Derek Rayner, (later Sir Derek Rayner), a Director of Marks and Spencer, was brought onto the Procurement Executive Board. (In exchange, a general was put onto the M&S Board.). Rayner's remit was free ranging. He held regular meetings to discuss progress of the ongoing programmes and seek subjects for new studies. They included topics such as -the organisation of military hospitals in Germany in the event of war in Europe- the introduction of contractor operation of large facilities run in government and/or the transfer of them to the most appropriate industry. Although the Air Establishments had many large facilities, built originally to serve a multi-company industry, but which, through company amalgamations were now reduced to one company, they were not the most obvious targets for change. The Navy and the Army had extensive design and development facilities as well as a manufacturing capability. Simultaneously, there were early signs that MoD wished to transfer as much as possible of the project risk to the contractor by tighter contract specification.

My Next Move

Changes were taking place in the Director General, Engines area in the Procurement Executive. Bob Weare was retiring and it was proposed that Ivor Davidson, Director of NGTE, was to move to the DG Eng post in London. I was lucky enough to be appointed to the Director NGTE post, the first non-engine specialist to be appointed since its first Director, Dr. Roxbee-Cox (later Lord Kings-Norton). One of the tasks for me was to implement the proposals to integrate the naval engineering establishments into NGTE.

Chapter 9

Director of the National Gas Turbine Establishment, 1974-1980.

The move to NGTE was very welcome. Not only was it a super job but also it moved me nearer to home without that train journey to London. Long ago, Alfred the Great bequeathed Pyestock to his nephew Ethelm as part of the Hundred of Crondall. Later the Monastery of Winchester administered the land. By the mid-1800's, it became part of a military training area. The NGTE site covered 235 acres and included part of the Bramshot Golf Club that was commandeered during WW11. (As a gesture to its history, the twelfth tee was fenced off and preserved.) In 1944, a new company was created by combining the Research and Development part of Power Jets Ltd under the leadership of Frank Whittle, the inventor of the gas turbine engine, and the RAE Engine Department led by Dr. A. A. Griffith. The Chairman of the new nationalised company, Power Jets (Research and Development Ltd) was Dr. Roxbee-Cox (later Lord Kings-Norton).

In 1946, this company was brought into the Ministry of Supply as the National Gas Turbine Establishment under Dr. Roxbee-Cox as Director. At that time the Establishment was operating at Pyestock and Whetstone. During the 1950s, under Hayne Constant, the site at Pyestock was enlarged and the development of major engine test facilities was started to facilitate the testing, on the ground, of the gas turbine in simulated altitude conditions that represented, in every way possible, the conditions that the engine would meet in flight. This simulation provided greater safety, considerable economy and facilities that could be used by many different engine companies.

The Facilities at Pyestock

We derived considerable benefit from having, in progress simultaneously on the same site, a full range of work from fundamental studies at model scale at one extreme, to development running of the most powerful engines that were being manufactured at the other. Close liaison was also maintained with the government and industry in Europe and in the USA at the Air Force AEDC and the Navy at NAPTC. The Establishment mandate covered the exploitation of the gas turbine engine on land, at sea and in the air. It employed some 1200 people and had a high reputation for the research and the engine testing work that it did in the very extensive facilities.

During the first three weeks, I toured the site talking in detail to as many of the staff as possible and getting an excellent feel for the work being undertaken, the aspirations of the staff, their priorities and concerns. I noticed that not much mail was arriving in my office but assumed that my deputies were relieving my load until I settled in. At the time, Rolls Royce was still not operating as a fully private company. They were still RR1971 when they went bankrupt over the commercial aspects of the RB211 and the failure of the fan blades to reach an acceptable standard on the tests involving the ingestion of simulated 2 lb birds. I visited Rolls Royce and several of the other equipment companies and then settled into my office to undertake a considerable amount of reading to be up to date on the recent activities. After a few more days, it was obvious that the amount of paperwork coming in my office was still negligible, so I called in my two deputies, Mike Neale and Walter Fletcher, both chaps of great reputation, and explained my dilemma. They looked at each other and with a smile said "It has been our policy for some time, to keep as much as possible out of the Directors office". After a long and very pleasant discussion, we agreed a mode of operation that gave me all the information and control that I wanted and made their role and responsibilities clear. It also allowed me time to think about the planning for the implementation of the integration of the Naval Engineering Establishments into, firstly the management of the Establishment: secondly, the transfer of staff and thirdly, the equipment and building implications of the moves.

The R. & D. Programmes

The gas turbine is a remarkable engine. Its power to weight ratio is very high, the pressures, temperatures and stresses at which it operates are always close to the limits of the strength of the material. Its main application remains the power for aircraft, ships, power stations and pumping stations. (The RR Avon is used extensively for pumping oil and gas from the North Sea).

Mike Neale, as DD R & D, was responsible for six internal research departments and for the management of extra-mural research. He was also responsible for the research funded by Department of Industry. One of the key functions was the provision of technical advice to government on matters of engine technology.

Extra-Mural Research Division, under Les Airey, was a very important part of the NGTE responsibilities for gas turbines in the UK. It was unusual in that the ratio of spend on extra-mural to intra-mural research was the highest in the defence area, three to one. It required the closest collaboration between the intra and extra-mural programmes and even the most junior staff were brought face to face with the engineering problems of industry. Budgetary pressures made an ever-closer integration of our intra-mural and the extra-mural research programmes in industry an essential part of our strategy.

To make the combined Establishment and industry programmes as effective as possible, considerable attention and effort was required from Mike Neale. He chaired all the Research

Programme meetings with industry, which were held quarterly, and I managed to attend some 90% of them. There were many conflicting pressures. On the one hand, we had to ensure that the in-house expertise was maintained at a level sufficient to fulfil our advisory role to government and on the other, to move the technology forward as fast as the combined intra- and extra-mural resources would allow. In the early phases, two or three different approaches may be considered to try to achieve the research goals. With progress, it would become clear that some ideas would appear to have a higher chance of success than others. At that stage, some ideas must be dropped and perhaps only two continued. Where the work was done depended on where the most relevant facilities were located, where the expertise was, and whether the company was prepared to add their funds to enhance the effort available on any particular topic. Whilst working so closely with industry, the independence of the views of NGTE was very important. Despite many difficulties, most of the time, it worked very well.

Performance and Project Assessment Division, under Frank Armstrong, played an important role in establishing the targets and priorities for future research areas and its co-ordinating role is emphasised. The gas turbine is a complicated machine and a change in one parameter can have repercussions on several others.

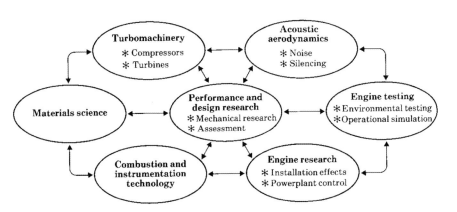

The Performance and Design Research Department and its Relationship with Other Research Departments and Engine Testing.

Advanced computing techniques were used to resolve the many possible permutations.

Materials Science, under Maurice Cockroft, was always one of the key areas in gas turbine advancement. Most engine failures are in some sense material failures caused by high operating temperatures, oxidation, corrosion, cooling air failure etc. In the 1970s, at the high temperature end of the engine, the strength of the nickel alloys could be improved by modifying the metallic structure. NGTE made considerable progress on the unidirectional solidification (UDS) process. As distinct from the conventional casting technique where molten metal was poured into a mould and solidified in a random manner with a granular structure. (left) The UDS technique required the mould to be surrounded by a heated furnace with a heat sink at its base. The latter initiates the crystal columnar growth in a controlled direction and, as the furnace is withdrawn, the crystals grew with a controlled longitudinal orientation. (center) The avoidance of transverse grain boundaries by UDS removes one source of weakness in the alloys and produces a creep/rupture ductility gain giving about three times the flight life at the same temperature.

Conventionally Cast, Uni-Directionally Solidified and Single Crystal Turbine Blades.

The blade characteristics could be further improved if a single crystal source in the root could be grown into a blade of only that single crystal (right).

Engine Research, including Installation Aerodynamics, was run by John Dunham. All installations require efficient intake and exhaust systems. The high levels of propulsion efficiency available from the use of high by-pass turbine fans could be seriously eroded by losses due to the installation. Of particular concern, on short cowl designs, was the gas generator after-body where, in the cruise at high subsonic Mach No., the flow from the annular fan nozzle was supersonic. Losses on this surface resulted from both shock waves and boundary layer growth and separation.

Theoretical methods were used to design after-body profiles that suppress the formation of shock waves and model techniques give accurate comparisons between the most effective ones. The RB-211-22b, with a simple 15 degrees conical after-body, achieved an improvement in flight of between 3.5 and 5.5% in specific fuel consumption.

Turbo-machinery Research was run by David Smith. Since the useful output of a typical gas turbine is considerably less than the power transferred internally from the turbine to the compressor, the achievement of good turbo-machinery efficiencies was and is of paramount importance. In a future of rising fuel prices and concern to secure the maximum effectiveness from military weapon systems, the turbo machinery stage loading and flow Mach numbers were progressively raised to achieve increases in cycle pressure ratio and by-pass ratio and minimise the number of stages required. Much of the research was aimed at reducing the losses in efficiency associated with high blade loading.

Acoustic aerodynamics was run by Martin Cox. The reduction of engine noise constituted one of the key problems facing the aviation world. Although in the early days, the high velocity gas jet formed the dominant noise source of the turbo-jet engine, the efficient high by-pass turbofan engine represented a relatively balanced assembly of noise sources. Apart from the jet, other noise sources were the large transonic fan and the turbines. Despite the progress that had already been made, there was still a considerable need for more information on the effect of flight in modifying the propagation characteristics, and in some cases, the actual generation mechanisms of important noise components. To this end, the large anechoic chamber was built. In addition, research on the acoustic treatment of engine intake and exhaust ducts for silencing purposes was undertaken in the Noise Absorber Facility. Sound pressure levels from 140 to 160 db could be produced and by placing the noise generators at either end of the duct, tests could be made with sound propagating with or against the main airflow. As the propagation and absorption of sound can be strongly influenced by velocity gradients, provision was made for boundary layer suction.

Combustion Research was run by John Macfarlane. In the gas turbine, the combustor is required to raise the temperature of large quantities of air through 800 degrees Celsius or higher within as small a volume as possible. This temperature rise must be accomplished with minimum loss of pressure energy and without creating "hotspots" in the combustion chamber or in the flow entering the turbine downstream. Generally, research was focused on the problems of cooling the chamber walls, maintaining good efficiency and a smoke-free exhaust with adequate burning stability at altitude conditions. At the time, environmental considerations imposed precise and severe limits on the composition of exhaust emissions. Accurate instruments and techniques had to be developed to measure the quantities of un-burnt hydrocarbons, oxides of nitrogen, carbon monoxides and smoke in engine exhaust systems. A mobile laboratory consisting of multi-orifice rapid quench sampling probe and a motor vehicle housing on-line chemical analysis equipment was built for "on the spot" monitoring of exhaust emissions at the engine test beds anywhere in the UK and comparing the standards of the test equipment across the western world.

Amongst the team, was a naval scientist who had a passion for combustion problems. The navy burns the most terrible heavy diesel in their ships gas turbines. It is often in the same tanks as seawater and has considerable biological growth at the interface. He invented a combustion can called the "pepper pot" because of the graduated spiral pattern of holes through which air was fed into the combustion chamber. The effect it had was to greatly increase the efficiency of combustion and particularly, decrease the particulates in the smoke as well as its infrared visibility. This change would have considerably reduced the signature of the ship detectable by modern sensors and reduce its vulnerability to attack. We were unable to persuade the Ship Department that it was worth greater investment with the view to installation in naval gas turbines. We never really knew why- it could have been the "not invented here" factor!!

Power Plant Control. The provision of an accurate and responsive control system is crucial to the performance of an installed engine. The control system for a complex military by-pass engines with reheat, in which several variables must be controlled simultaneously, can add considerably to the weight of the total propulsion system. At the time, the fuel control systems could best be described as "a plumber's nightmare". They were extremely difficult to modify to meet any new engine requirements. NGTE was doing research into lighter, cheaper and more reliable systems including work on system concepts and logic, control hardware (sensors, actuators, etc.) for a digital power-plant control. Many of the building blocks had been funded via the extramural research programme in the engine control industry and we had successful completed of a series of engine tests using a laboratory digital computer. As we saw it, the next step was for industry to produce a flight-worthy, full authority digital control system for test bed and flight evaluation that would demonstrate the value of the concept.

Although our budget had been sufficient to fund the research programme, I could not see that it would be possible to commit more than £200k in any year to it. What was needed was funding for a demonstrator programme costing in the region of £3m to £5m over two years. Such a programme was the preserve of the D G Eng. branch in H.Q. The case that would generate that sort of funding had to be made. NGTE offered to fund the first year at £200k to get the programme started provided that DG Eng would pick up and fund the remainder of the programme. Industry was pressing us hard to get started. I told them that we were prepared go ahead but I needed the assurance from H.Q. before committing. When Industry went to London, D.G.Eng would tell them that it was up to

NGTE to start the programme. Ultimately, I had to involve John Charnley, CER, and Sir Douglas Low, C.A.

A meeting under their joint chairmanship was arranged in London. Eric Lewis spoke for D.G.Eng. giving his support to the programme. I had decided that this meeting was the opportunity to tell our respective bosses, CER and CA, how important the programme was and why the funding had to be found. The presentation lasted about 45 minutes. I finished with a summary of the importance of the work and that we could, with funding from the research budget, start the programme. I also pointed out the dangers, particularly if we went for a competitive industrial programme, as the grouping would be decided by a small research contract. C.A.'s reaction was immediate. He reminded me that industrial grouping was a H.Q. responsibility and H.Q. should take over the management and funding. It was a wonderful result. The budget was found and we did not even have to commit the £200k. It was quickly absorbed by other important research objectives.

Not long afterwards, Ivor took early retirement and Mike Neale was promoted to be D.G.Eng.. Our relationship with H.Q. improved dramatically.

The Engine Test Facilities.

To test aero-engines in simulated free flight conditions requires vast quantities of air to be moved around the site at low and high pressures. To do that requires a great deal of electrical power. The maximum load that we were able to take from the mains was 105 mega watts and we added to it 20 megawatts from our own power station. 125 megawatts is enough electricity to power a medium sized town. The 20 megawatts was also enough to run the base load of NGTE and RAE when we were off-line from the grid. The price of electricity was negotiated every year. It was always at a discounted price (about half that paid by a domestic user) because we guaranteed to come off-line within three minutes of the Electricity Board making that request. This situation was most likely to arise in winter when the domestic load increased rapidly. The facilities for testing engines were enormous consisting basically of a centralised air supply/exhaust extraction plant, conditioning equipment for drying, heating or cooling air, connected to five test cells with their data acquisition systems and control rooms and a data processing centre. There was also a sea-level test bed used for calibrating engines before installation in the altitude cells and for research purposes.

Cell 1 was originally designed for the free jet testing of ramjets and was later used for testing model air intakes for supersonic aircraft.

Cell 2 was used for "connected" tests. (The test air is directly connected to the test engine or reheat system). It could cover a wide range of jet engine conditions at low altitude. In the case of the RB199 reheat system, most of the tests were done using Cell 2 as an engine simulator. It was able to test the mechanical integrity and combustion performance over a wide range of flight conditions. In fact, it was the only cell in Europe where the conditions in the reheat chamber at a flight Mach No. of 1.2 at sea level could be simulated.

Cell 3 was large enough to do "connected" tests on engines such as the Olympus 593 (Concorde), Pegasus and Spey engines. The mechanical integrity of the engine was always checked and a series of cycles of power settings (often up to 700) between flight idle and the maximum continuous thrust rating were completed.

Cell 4 was a large "free-jet" supersonic facility with a variable Mach No. working section and with the flexibility to vary the incidence and/or yaw condition whilst running. It was able to test a Concorde intake and an Olympus 593 engine over a Mach number range of M 1.7 to 2.3 and pitch angles of -2 to +10 degrees. The intake, engine and nozzle control systems could be tested and their behaviour examined when subject to various simulated failures such as: - jack failures on the intake ramp and the simultaneous failure of an adjacent pair of engines, an event, which if it occurred in flight, would result in a rapid yaw transient.

Cell 3 West was built for "connected" testing of high by-pass ratio turbofans of the 50,000 + lb thrust class such as the RB 211. Engines could be tested, in safety, through a wide range of flight conditions. To achieve the highest efficiency from an engine, it is generally to operate close to the "surge" boundary. In flight, engine surge is a terrifying experience. With a series of very loud bangs, the flow through the engine actually reverses and flames come out of the front. Quite apart from problem that the front is supposed to be the cool part of the engine, the extreme variations in pressure can readily induce structural failure of the ramps or intake vents as well as the engine itself. Whilst testing the RB-211,

850 engine surges were induced to identify the conditions under which they might arise. The engine survived all these tests, which gave great confidence in its robustness and design. In addition, the measurement of thrust had to be made with the highest possible accuracy as the results were used to help settle disputes between the airframe and engine companies on whether it was the engines that were not producing the guaranteed performance or the airframe that was producing more than the predicted drag.

Engine Icing Trials of the RB-211

Icing trials could also be undertaken. The three types most often in the programme were: -

i) icing of the engine, particularly the fan. It was important to see, not only the way that the ice accumulated on the fixed and rotating parts of the engine, but also the way that it broke off and the reaction of the engine to ingesting large lumps.

ii) icing of helicopters and engines and the shedding of ice from the surfaces. Major elements of the engine/fuselage combination were used during icing trials.

Icing Trials. Sea King Helicopter.

iii) icing of helicopter rotor blades and the effectiveness of various blade heating systems for the prevention of ice accretion.

The Naval Marine Wing

RR Marine Olympus TM38 on Test.

The Naval Marine Wing was a separate department and undertook research and development of RN gas turbines for Director General Ships.(DGS). In 1967, the RN decided that all its future surface ships would be powered by gas turbines. Marine gas turbines are based on aero engines incorporating an additional free power turbine. Some redesign was required to make them resistant to the corrosive environment at sea level, burn dieso instead of kerosene and be resistant to the shock loading from underwater explosions. At Pyestock, extensive endurance trials were undertaken in the NMW. Engines were installed in a ship module with ship ventilation, shock mountings, control system and vertical intake sections and with two dynamometers capable of absorbing 40,000hp normally taken by the ships propeller system. Cyclic endurance running was undertaken with throttle movements

Gas Turbine Powered Ship at Sea.

Pitting of Marine Turbine Blades due to Salt Corrosion and Carbon Particles Embedded in a Corrosion Pit-x600.

as vigorous as those met in ship usage. The simulated marine environment was achieved by injecting seawater into air and fuel to give conditions of 0.01 ppm NaCl in air and 0.3 ppm Na in fuel. The testing included performance measurement, compressor cleaning studies, maintenance evaluation, vibration checks and health monitoring.

The ingestion of salt at sea is a major problem. An important element of the work of the NMW was to evolve a filtration system that reduces the salt ingestion to an acceptable limit of less than 0.1 ppm NaCl under all conditions of sea spray and yet is compact, light, inexpensive and has a low-pressure loss. Without it, compressor fouling results in lost performance and sulphidation attack on hot end components results in pitting and then component failure.

It was the excellent working relationship of this group with DG Ships that did much to ease the transfer of the three DG Ships laboratories to NGTE in the rationalisation programme (discussed later).

Management and Industrial Relations

NGTE was fairly unique in the MoD in having such large facilities. It was vital that they be used to maximum efficiency and an agreement had been negotiated with the staff and trade unions for operating a basic two-shift system with no overtime. It operated from about 0700 to 2300hrs with a silent shift during which some maintenance was undertaken. It allowed us to start testing at 0800hrs and run the facilities until 2230hrs. This scheme minimised the disturbance to the local population particularly in the nearby suburbs of Fleet. By 1978, trade union power seemed to be well on its way to ruining Great Britain. These problems were reflected in the Civil Service where the government was allowing the public employees pay to rise much more slowly than that in industry, resulting in staff shortages in those areas where significantly higher pay was available in industry. We suffered from a serious shortage of electricians and, in the climate of exercising their muscle, the Electrical Trades Union decided to go on a series of "work to rule" lightning strikes. It caused massive disruption to the test programmes.

I had a lot of sympathy with them. For a period of six months, I was being paid less than my deputies. However, it was not a situation that I could allow to continue. Much to the alarm of my management board and the Industrial Relations side of MoD, I decided to call the whole establishment together, in three sessions, and brief them on our current and future programme, the impact of the strikes on the operations and the possible implications for the future. I also told them that I had decided to suspend the Industrial Whitley Council and deal with each of the trade unions separately. I would also be writing to the T.U. District Branches and HQ and seek their assistance in sorting out the mess. The presentations helped everyone to understand the problems and I had considerable support within the establishment. Someone wanted to send me a message and in the snow outside my office window trampled out the words, "Ignore the strikes and smile. God loves you." Throughout these problems, the personal relationships were good and normal working was restored quite quickly.

During this period of unrest, a Minister, Bill Rogers, visited us. After taking him around the site and explaining the work we were doing, it was my policy that, at some time during the day, I would introduce him to the Staff and Trade Union representatives. Having introduced him to them all, he stood in front of them and said " I understand that all is well with you and the only problems you have are with your Director." At that point he bound off up the stairs towards my office leaving the representatives with their jaws dropped and me, somewhat surprised, giving chase.

At about the same time, the District Manager of the National Westminster Bank invited me to a luncheon at which the national problems were discussed. I can remember the shock on the faces round the table when I was asked; " Did I not think that the current unemployment level at 5% was excessive?" and I replied "It did not seem to matter much. I was too busy trying to keep the other 95% working!"

We had problems between our own and the RR unions. At RR Bristol, they had what was called the "Shop 4" problem. The RR engineers wanted to do all the work on an engine. The NGTE engineers felt that it was entirely their job to install the engine in the test cells and attend to its interface problems with the test cells throughout the trials. I fully supported my engineers. RR engineers would do all work on the actual engine. However, the RR trade unions insisted on being locally at Farnborough throughout the test programme. In fact, they stayed in the Queens Hotel and played cards for that period. Eventually the problem was "bought out" as a better balance was achieved between management and T.U. power.

The Testing of the FJR710 for Japan.

I was keen to take on other work for the test cells by encouraging other countries to test their engines at NGTE.

The Japanese government was supporting the development of its aerospace industry. They had designed and built the FJR710 aero engine and wanted to test it in an altitude facility. They went to the USA and France to negotiate a testing programme. Then they came to us as part of the competition. We showed them the facilities and then discussed the detail of the programme and, in particular, how the work would be done. Because of the great distances involved, we wanted two engines delivered at the start of the tests. This arrangement would guarantee that the tests could be continued even if we had trouble on one engine. I left it to Walter Fletcher and his team to reach an agreement and then to call me back when it had been achieved. I asked the Japanese to describe to me what they understood about the techniques and procedures that we would be using. The description that they gave worried me because there was no discretion or flexibility in the programme. The agreement was re-negotiated and with some misgivings on their part about what their contracts and accounts people would say, the agreement was signed.

On the first day of the tests, it was clear that the oil cooling system on the engine was inadequate at altitude. The Japanese were greatly dismayed as they thought that it was all over. The ingenuity and flexibility of the Cell 3W staff then came to the fore. The Pyestock engineers rescheduled the programme to do the low altitude tests first where the cooler was adequate and simultaneously devised a separate oil cooler for connection to the engine to produce the correct oil temperature for the high altitude tests. By that means we were able to run through the whole performance programme of the engine. The company could then redesign and enlarge the oil cooler back in Japan. They were "over the moon". The trials were completed on time and the Japanese engineers were getting their annual bonuses for their achievement.

The Chinese at Pyestock.

During the mid-seventies, China was emerging from the Cultural Revolution, a period of communism at its most extreme and oppressive with isolation and political unrest. Its relationship with Russia had been abandoned. Eventually the Chinese Government decided that to "modernise" it had to become involved with the West. It had been negotiating a partnership with RR to build the Spey engine under license and be taught the techniques and skills of manufacturing it. During these visits to R.R., they also visited Pyestock, became interested in the engine test facilities and wanted to learn how to build engine altitude test facilities.

The Chinese at Pyestock.

In 1977, the U.K. agreed that a team of about 20 engineers from China, led by Professor Kang Yi of the Peking Aeronautical Institute, should visit Pyestock for three weeks to be briefed on the types of engine testing being done and the facilities that would be suitable for testing modern gas turbine engines at sea level and at altitude. Each day they were given a specialist brief on a particular facility and were able to have discussions with the experts of that area. They were ravenous for information, very widely read on the subject but, when the discussion moved from the theory to the practical means of doing the work, they had some difficulty in understanding the issues involved. At the weekends, we took them out either in small groups or all together to see something of our country and the social and sporting interests of both the Establishment and the area. The local press suggested that Russia would be in a bit of a flap about the activities behind the security fence at Pyestock. The visits were highly successful and did much to help the sales and collaboration programme with RR. The Chinese wanted Pyestock to be their agent for the design and erection of engine test facilities in China. However, it was not appropriate for a government establishment to fulfil that role but it was agreed that we would act as their technical advisers.

In June 1977, I received an invitation to visit China. I applied to MoD for permission to accept the invitation and, on 16th April 1978, a team of three of us Austin Seed, Martin Holmes and I, with Evelyn, entered China by train from Hong Kong. After giving up our passports to Mr.Yu, a Foreign Affairs Officer, we continued by train to Guangzhou (Canton) and flew by China Air to Beijing. There Professor Kang Yi and a team from the China Aeronautical Society met us. We were housed in the Peking Hotel, full of foreign guests. In the elevators, it was noticeable that the 6th floor was missing. We assumed that this floor was a centre where the listening devices placed in the bedrooms was received. None of the Chinese ever came into the suites put at our disposal even though invited for a cup of tea.

The first morning was occupied with discussions on the programme for the visit. We had each prepared a paper and lecture to an audience of invited engineers and scientists. We were to give the lectures simultaneously in three different conference rooms. Unfortunately, only one projector was available. I contacted the Military Attache at the British Embassy and arranged a meeting for the afternoon. He and the Commercial Counsellor were able to loan us the necessary equipment.

The lectures were to be followed by six sessions on each of the next six mornings of discussion periods of 3 ½ hours. My audience was from the Shenyang Engine Factory, the Hsian Engine Factory and the Peking Institute of Aeronautical Engineering. Each attendee had received a copy of the lecture two weeks earlier and they were well read in the latest UK, European, USA and Russian papers and used them to illustrate their questions. Unfortunately their experience was too limited for them to be able to pick out the important issues in a subject and identify for themselves where their work could be improved. The group seemed to have no concept that an aircraft would be designed to do a particular task and the test programme would need to be devised to establish that the performance requirements were met and at the flight conditions to be expected. The long periods of discussion exposed areas where detailed explanations would be welcome. Each evening, the three of us discussed the issues raised during the day and often prepared additional 30minute lectures on the topics raised during questions to give more detailed explanations.

Most afternoons and evenings, some visits of interest or official dinners were arranged. On the free evenings, we took the opportunity to accept an invitation to visit the British Embassy, to work or to stroll around Beijing. On the visits to the Palaces, Operas etc. we were treated regally. We were given immediate viewing of any exhibit no matter how crowded the exhibition. When the three of us were working, a very pleasant young lady called Miss Wong looked after Evelyn. She took her on visits to schools, hospitals and places of interest.

Whilst in Beijing, we visited the Aeronautical Institute and saw their teaching and test facilities. We gained the impression that the Institute staff had not considered carefully the real needs of the aircraft and engine industry for designers and research and development engineers. This lack of appreciation of the magnitude of the effort required to make the large leap forward that they planned, probably reflected the position of industry in China that was still centred on the building of foreign aircraft and engines (mostly Russian) under licence.

We were next to travel overnight by train to Shenyang, north of North Korea. Our party of seven was taken to Peking Railway station in the evening by a group of cars. On arrival, our luggage was all loaded on a handcart pushed by one small old lady. Evelyn protested to me that I must not let it happen. I had to take the view that this was the system and the "little old lady" might be insulted if I objected as it might imply that she could not do her job. We were then taken into the great hall of the station that had two escalators at the far end. The station was extremely crowded with people sitting, moving and sleeping in every possible position. As we approached the escalators, they were cleared until my whole party had used them. It was very surprising and impressive.

The rail journey was a pleasant one. Waking up early in the morning, it was surprising to see the peasants working the fields at dawn, having walked from the scattered small villages with no more than tracks connecting them. In Shenyang we were accommodated in the Liaoning Hotel in the suite used by Chou en Li. I even slept in his bed.

Our guides said that there would be insufficient time to see both engine factory and its test facilities and the aircraft factory and we were given the choice. Having seen the engine facilities in Peking, much to their surprise I opted for the aircraft factory. There was a great deal of discussion and shuffling of feet but eventually they announced that they had managed to arrange both. At the aircraft factory, which they said that they had especially opened on their day off for us to see, they were producing Russian Mig 19 fighters at a rate of 400 per year in the ratio 50/50 fighters /trainers. They worked two shifts, six days per week and 10,000 workers were employed, including about 1000 engineers. Nowhere did we see a drawing. It appeared that the Russians had taken all drawings when they left. Everything was being manufactured from metal templates. In one area where they were building a cockpit, one man was in the cockpit doing the work. Around the cockpit were about five others, all watching. We were told that they were apprentices, all learning on the job.

The engine factory also employed some 10,000 people, producing about 1000 Jet 6 engines each year for the Mig.19. The lost wax process was used to produce turbine blades. Little attention seemed to be paid to quality control. In the engine build workshop, we were able to closely examine the compressor blades, which had poor finish as well as obvious variation in blade angle and trailing edge thickness. On final test, the engines were run up to idle in an open workshop with the exhaust gasses directed out of an open window. If this test was successful, the engine was installed in a test bed and run for a total of 17 minutes, seven of which were using reheat.

The sole function of the test facilities at Shenyang, which were quite extensive, was to assist in trouble shooting problems in production. There was no forward thinking on the type of programme required. They were requested to do work by the HQ in Beijing and reported to them when it was finished. They then seemed to disappear home to work on their vegetable production, which they could sell, until the next request was received from Beijing.

After Shenyang, the working part of the visit was over. We had attended many official dinners and made many speeches. They were an opportunity to exchange greetings and gifts. The dinner parties were lavish and very cordial. Any joke at the expense of the Russians went down very well. At other times, because the sense of humour was rather different, the prefix that the next comment was to be a joke was always a guarantee that everyone would fall about laughing. We visited Shanghai and had a trip on the Yangtze. We then visited Hangchow, a beautiful holiday resort on a lake. We visited many historic sites and temples, silk factories and palaces.

We then flew back to Guangzhou for final discussions and official dinner. Everywhere we had been in China, the children had greeted us like old friends, in schools, in art centres for music and dance, in games training facilities and in the national parks. Whilst being escorted back to the border, I asked our escort why the children were so friendly? I explained that they must know that we were the long noses, had taken part in the opium wars and historically had not always been friendly. He said that it was natural but I pressed him on the point. I asked him what they were taught in the schools? He replied that they were taught to be good communists, to work hard, help each other and love our foreign friends. I asked him how they knew who were their foreign friends? He replied, " If you are here, you are a foreign friend!"

The visit to China was a fantastic experience. I found it impossible to believe that the Cultural Revolution had been so self-destructive. In the communes, they even destroyed their own crops and ate grass to demonstrate their revolutionary fervour. As the "gang of four" had just been put in prison, everyone was keen to talk about the deprivations of that period. The tortuous language of the official press had to be experienced to be believed. The people we met were delightful but the overall political system and organisation left a great deal to be desired.

It was not long before other Establishments, RAE and RSRE were invited to China. We were proud to advise them on what to expect.

Rationalisation.

In March 1976, the Defence White Paper announced that the rationalisation proposals by CER should be pursued at all speed. We proposed that the bulk of the naval engineering work should be integrated into a Naval Mechanical Engineering Department at Pyestock under Captain Hurworth, then Superintendent of AEL at West Drayton. He became Deputy Director (Naval) responsible to me for the task of integration and equipment and staff moves to NGTE. The Admiralty Oil Laboratory became the Petroleum, Chemistry and Technology Department,- The Admiralty Marine Establishment became the Naval Auxiliary Machinery Department - the Admiralty Engineering Laboratory became the Naval Engineering Department and the testing of naval gas turbines became the Naval Marine Wing and was integrated into it.

The move of the staff and laboratories from AOL at Cobham was the easiest to arrange. Buildings into which they could put their engines and equipment were largely available and thus it was an issue of arranging the earliest convenient time, considering the programmes already under way.

Making plans for physical moves of large equipment from West Drayton was always difficult. The demands of the service programmes delayed some of the proposed moves. After arranging for the moves of the large eight batteries of the DC station, in 1978, the new SSK submarine was agreed and the availability of the battery station as the major DC electrical facility, vital for the programme, delayed any possibility of a move.

In parallel, it was important that the staff began to feel that they were all part of one establishment. To this end, I decided that as posts became available within the establishment that they would be advertised first within the establishment so that anyone who wanted to level transfer had a chance to do so before the posts were advertised more widely for level transfers and for promotion. The scheme was not entirely approved by HQ but it seemed to me that it was attractive and could achieve the desired results. By this means, staff began to transfer from West Drayton to Pyestock. One day, at a Board meeting, after discussing the programmes and rationalisation position of the Naval activities, I said that I thought that the advertisement of posts in NGTE was succeeding and we were getting quite a few moves within the Establishment. Captain Hurworth, with a very white face and sombre look agreed that we are certainly getting moves but his ex-staff were not happy. I asked why? He replied that in the drawing office they were charging the new staff 1p per cup more for their tea than the long serving staff at Pyestock. When I stopped laughing, I agreed to investigate. After the meeting I walked down to the Drawing Office and with the Head of the Office walked around the chaps and discussed their work and how they were getting on. When we got back to his office, as there had been no comments I asked him about the tea "swindle". He said that, as in the Civil Service things like tea had to be paid for by the individual, the tea had been bought in bulk at a time when it was much cheaper than current price. The chaps had felt that the new comers should pay one penny extra per cup until the current supplies had been exhausted. I still found it very amusing but that was democracy at work.

AMEE at Haslar was a different problem. I made regular visits there to make them feel part of our activity but the possibility of much movement of staff or facilities and of saving was far from clear. Some of their most important work using very high pressure steam for testing relief valves depended upon the proximity of ample supplies of water. On one occasion, the Captain wanted me to watch a typical trial. We watched from the observation chamber whilst the super heated steam was brought up to 4000 psi for the test. Super-heated steam was jetting out of the numerous joints in the piping and men with sledge hammers and spanners with handles about 4 ft long were hitting the pipes with the hammers and simultaneously trying to tighten the screw joints. I asked for the trial to be stopped and for the management team to come with me to the Captain's office. When there, I asked them what was going on because I had not seen anything so dangerous for a long time. They did not know and commented that they had not seen it before. I left instructions that in future all adjustments should be made when the steam pressure had been dropped and that TV cameras should be fitted in the control room so that any steam escape could be seen without anyone being in the laboratory. I also wanted an explanation of the day's demonstration. Two days later, the Captain telephoned me to tell me that the chaps had decided to put in a claim for danger pay. They wanted me to see the risky nature of their work before they put in the claim. I thought that the demonstration was very effective but perhaps disappointed them because I chose the route to safety.

AMEE was located in the middle of AMTE (the Admiralty Marine Technology Establishment) and all sorts of minor but important niggles began to occur between the two establishments. As the detailed work on the moves of staff and equipment showed little saving, I persuaded DG Ships that the work for him as a customer could be equally well done under AMTE and AMEE should be transferred to it. AMTE was by this time under the Admiralty Research Laboratory at Teddington and I agreed with the Director of that Establishment that we should do a joint presentation to the staff explaining the plan. The presentations went well. I think that everyone accepted that the new arrangements would be better. After the presentations, I went into the social club bar and was soon in discussion with the Chairman of the Shop Stewards Committee who was an ex-seaman, pleasant, broad and tough. He said- " I told them what it would be like. You would stand up there, smile at them and tell them how good the new scheme was going to be. They would nod and say Yes Sir. That is exactly as it happened."

The Noise Test Facility and the Visit of the Duke of Edinburgh.

Studies of the jet noise problem began in the early 1950s. At that time, jet engine noise was predominantly due to the high velocity hot gas in the jet as it mixed with the atmosphere.

155

As engine design evolved to fulfil the promise of the improved fuel economy offered by the ducted fan cycle, turbo-machinery noise increased in importance because this tonal property is part of the noise spectrum that people dislike. This varied from the band-saw noise of the fan to high-pitched whines from the turbine. A series of psycho-acoustic experiments involving many thousands of subjects were carried out to provide scientific consolidation of the Perceived Noise Decibel scales and the NGTE results played their part in forming the criteria subsequently adopted in internationally agreed legislation.

In the late 1960's, there was increasing pressure for noise reduction on amenity grounds and it was apparent that the proposal to introduce International Noise Certification for aircraft by the early 70's could seriously prejudice the aero engine industry of any nation not able to meet the exacting standards that were likely to be imposed. In the UK, government funding in the noise research area was trebled between 1965 and 1969. Towards the end of this period, discussions were held with Rolls Royce and other interested parties to review the existing facilities and make proposals for new facilities to meet the requirements for noise testing. To cut a long story short, it was decided to locate the facility at NGTE because suitable air supplies were available and construction elsewhere would have entailed substantial extra cost.

In April 1969, a proposal was submitted by NGTE and approved by the Treasury in March 1970 at an estimated cost of £2.2m. The feasibility stage showed that the initial concept of a large single multi-purpose building was not practicable and that there were considerable advantages in separating off the Absorber Facility. It could be constructed more quickly and used to meet the urgent need for design data on absorbent materials. The revised plans were approved in 1971 and the Absorber Facility completed in August 1972. The large anechoic chamber was completed in April 1974 followed by the fitting of machinery and instrumentation.

Both facilities were paid for by the Department of Industry at £2m i.e.10% less than the original estimates. The main features of the NTF were the engine rig for testing turbines or co-axial jets both hot and cold.- the fuel used was butane gas. The nozzle was 18 in diameter, large enough for research purposes and able to test engines of the Viper class. The noise measuring equipment on pylons and traversing boom was able to make "far field" noise measurements, the microphones being 40 ft from the nozzle. The absorber lining was made up of 22,500 wedges to simulate a free space noise environment for frequencies above 80 cps. During the building of such a large facility, it would not be surprising to have had a number of accidents. In fact, there was only one in which a workman fell from an 18 ft scaffolding and landed on his feet without injury. The sound absorbent properties were so good that no one heard a word he said!

We felt that the opening of this facility was so important that our request for a visit of His Royal Highness, the Duke of Edinburgh was arranged for 11th June 1975 and the highlight of this visit was to open it. He was the first member of the Royal Family to visit the Establishment. (Mr. Harold Wilson had opened Cell 4 in 1967). The Noise Test Facility was especially illuminated and prepared for the opening ceremony. We arranged to have a rehearsal and Sir George Leitch, the Permanent Under Secretary, PUS, at the MoD came to run through it with us. On that day, the weather was hot and when the helicopter landed on the sports field in front of the dining room, where we were to have lunch, the windows were open and the dining tables were covered with newly mown grass. On the big day, the windows were kept shut until that danger was over.

The Main Chamber of the Noise Test Facility.

The Prince was due to arrive for lunch at 12.45 pm after opening the Bernard Sunley Activity Centre, part of the London Federation of Boys Clubs of which he was Patron. A press photographer had been upset him and it took some time for him to set the incident aside. Sir George Leitch and I met His Royal Highness at the helicopter, and we walked across the grass to the dining room where he was welcomed by Sir Ieuan Maddock, Chief Scientist at the Department of Industry and John Charnley (later Sir John), my senior staff and Mr. S. Baldwin, Chairman of the Staff Whitley and Mr. Don Barham, Secretary of the T.U.Whitley. I had been advised that for drinks, a gin and tonic

Meeting Prince Phillip From the Helicopter.

and a tomato juice should be offered. If he took the gin and tonic all was fine; if he took the tomato juice, then I was in trouble. He took the tomato juice. However, all went well and as the meal progressed everyone relaxed for a splendid afternoon.

The staff gave him an enthusiastic welcome. At the opening ceremony, Prince Phillip was so impressed by the building and the lighting arrangements that he made an impromptu speech and during the afternoon he was introduced to over 100 people. On the staff at Pyestock, we still had 27 people who had worked for Frank Whittle at Power Jets at Rugby or Whetstone. Many of them had served for about 30 years in the gas turbine field. The tour was completed on time and we all rated the visit a great success.

Having set up all the exhibitions, it was important to get full value from them. On June 12th and 13th, we were able to welcome parties from Rushmoor Borough Council and from Hart District Council. It was a great opportunity to show them the scale of the facilities and explain to them our good neighbour policies. The issues of greatest importance to them were: -

i) Noise- we spent large sums of money in containing the noise inevitably associated with our work. The standards we set ourselves were tighter than the existing codes of practice in both day and night time conditions. We welcomed communications from the public that enable us to identify the sources of annoyance and, whenever possible, we did our best to mitigate the effects. Each complaint was fully investigated and the complainant courteously informed of the results. We had received 12 noise complaints over the last 4 years and all had been satisfactorily resolved.

ii) Pollution. We had received 6 complaints in the last 4 years, all related to the smell of kerosene. They were all caused by tests in an engine failure mode involving "flameout" combined with abnormal wind and weather conditions at the time. As soon as we were aware of the complaints, the particular tests were stopped. Arrangements were made to obtain local weather forecasts prior to any test runs that might involve significant quantities of un-burnt fuel exhausting into the atmosphere so that this unfortunate combination could be avoided.

iii) Oily water. We ensured that all water used on site was thoroughly cleaned before discharge into Cove Brook. The Thames Conservancy checked the final effluent for acidity/alkalinity before release. It is sufficiently pure to re-circulate. In fact, we re-circulated 25 million gallons of water each year to conserve the natural resources. After removing oil and soot, the residue was taken from the site by a reputable cleansing contractor.

iv) Electricity. By stringent programming both within NGTE and RAE, we restricted our joint demand on the National Grid to the minimum compatible with our testing programme. We had the capability of generating a significant amount of electricity and at times of National Emergency could, at certain times, feed the National Grid.

v) Apprentice School. This school provided ideal opportunities for the local youth to train for skilled jobs. We had roughly five times as many applications as we had places.

There were all sorts of other ways in which we interacted with the local community. For example, we supplied 450 meals on wheels per week from the canteen, we had helped to restore the Basingstoke Canal, and our emergency services had excellent liaison with the local emergency services. In Hart, NGTE was also a substantial contributor to the local rates.

Finally, on Saturday 14th June, we had an open day for the workforce and their families so that they could really see what everyone did. As we had not had one for 9 years, it was a great success. (Having Sir Ieuan Maddock with us for the Royal visit, reminded me of the story of an event involving him that occurred some years later when Michael Heseltine became Minster for the Department of Trade and Industry. On the first day in office, he called all his senior staff into his office and after some informal chat said " Well gentlemen, by the time I was 25 I had made my first million. What have you done?" There was a stunned silence. Eventually Ieuan Maddock chirped up and said, " Minister, I don't know whether this counts—but I have detonated three atom bombs")

The Apprentice School

The Apprentice School under Mr.Hatto, was excellent. Though we had five times as many applicants as places, we always tried to take a spectrum of abilities into the school. The best would go on to university and graduate, the average would stay at Pyestock for five to ten years. Then financial pressures would induce them to seek more highly paid employment. Often they came back to us after they had reached 50+ years and stayed until retirement. Generally we offered jobs to about 95% of the apprentices. (It helped to maintain their enthusiasm through the course.) Every year there was a presentation of certificates to all who had passed their examinations and of 15 prizes when we could show our appreciation of those who had excelled.

Every year we invited some of the great and good to present the prizes and to talk to the apprentices and their parents. In the year that Prince Philip visited us, we invited Professor Ffowcs Williams of the Department of Engineering at Cambridge. He had worked closely with us on the studies of aircraft noise. Many other important celebrities came to present the Apprentice Prizes , Sir John Charnley, Mr Dennis (Joe) Lyons, Mr Jack Daniels (DG Ships), Dr J.W.Drinkwater, Ex-Whitworth Scholar and ex Deputy Director NGTE, Dennis Higton, ex graduate apprentice etc. All were chosen because they had very important messages for the apprentices. Each was presented with something symbolic of the Establishment, such as bookends, with stub wings and an RB211 engine installed below them, made by the apprentices as an exercise within their training.

One of the complaints of the Apprentice instructors was that the apprentices were unable to do simple arithmetic manipulations such as convert fractions to decimals and vice versa. I decided that I should write to the Head Masters and Head of Maths Departments in the schools in our catchment area and invite them to a meeting at Pyestock to discuss the problem. At the end of the meeting, they agreed to examine the syllabus to see whether traditional maths or new maths covered the subject properly and we would make a film to show the importance of mathematics in work. After three months we met again. We had produced the film but they had been unable to establish the reason for the problem. Nevertheless, we agreed that it was important to continue the search for the source of the problem and, if possible, find a solution. After further work we discovered that only 34% of the maths teachers in our catchment area had an "O" level in maths- i.e. 66% did not. Some 20 years later, I am sure that the examinations position has improved because the GCSE is now easier but the shortage of good maths and science teachers continues and will do so until the system pays the rate for these particular skills.

ISM Awards

When an industrial employee has served well in government work for more than 25 years, their names are submitted to MoD for the award of an Imperial Service Medal. The task of presenting these medals was delegated to the Directors of Research Establishments and for me it was both an honour and pleasure to perform those annual duties. As the wives had supported their husbands over all those years, we decided that as another exercise, the apprentices should manufacture small NGTE brooches to be presented to the wives and involve them in the ceremony. I was always impressed that they wore their medals and brooches so proudly.

Whittle and the Reactionaries

In 1944, the government imposed break-up of Whittle's Power Jets did not please him at all. He had a very loyal workforce who had helped and sustained him overcoming great difficulties through those early years. They objected to coming to Pyestock and Whittle felt that he could not bring himself to visit. They formed a group called the "Reactionaries". Their club badge was the rear view of the W1 engine with flames in the final nozzle, embroidered on a tie. With so many ex-Power Jet employees with us, there was an excellent opportunity to hold an annual Reactionary Reunion Dinner at Pyestock. Archie Simons was the Honorary Secretary and each year he put out a Newsletter updating everyone on the situation of their friends.

These reunions were always nights of stories of the past, of speeches and celebrations. Some members had become amateur magicians and entertained us with their skills. We were delighted that Sir Frank was able to come to one of these reunions. We showed him the facilities, which impressed him, and we always put on a display of historic pieces of equipment with their story. He was surprised to see a photocopy of his original notes with many examples of the early experimental machinery in the museum. Many of the exhibits had been presented to us by the families of those early pioneers who had unfortunately passed away. Much of it seems to have been lost or dispersed. For me, it was a great pleasure to be made an Honorary Member of the Reactionaries.

Health and Safety at Work

With the use of massive quantities of electricity, the movement of huge quantities of hot and cold air around the research rigs and engine test facilities as well as being insulated with massive quantities of asbestos, the health and safety of the whole work force required the constant attention of the management. Walter Fletcher had developed a safe key system of operating the facilities. Before anything could be run every key had to be in place. Simple things like the internal sweeping and surface finishing of the 8 ft diameter pipes around the establishment had to be done with the absolute confidence of everyone involved, particularly that those working inside the pipes, were safe. It had to be done regularly because any rust or dust blown through the engines would erode the turbo- machinery, blades etc. of the engines or research rigs on test, all of which were expensive and key equipments.

When the Health and Safety Act was passed by Parliament, it placed legal responsibilities, not only on management but also on the employees themselves. Many did not wish to accept them. Combined with the ongoing problems related to civil service pay and conditions, there was clearly great potential for disruption of the programmes. If we did not get the right man into the Health and Safety job, it could generate serious industrial problems for the Establishment. It was a very important job that required good contact with everyone in the workforce and with management.

I regularly made "walkabouts" alone around the Establishment and spoke to as many people as I could, discussing any problem that they had. I took it back to the management to consider and, if possible, solve. Often the issues were personal ones. I had a sandwich lunch in my office on one or two days each week. That was the opportunity for staff to see me alone for longer discussions. After one of these walkabouts, Don Barham, the Secretary of the TU side asked me if he could see me. We had a long discussion on the issues raised by the Health and Safety Act. He had been on several TU courses on the problems and was very knowledgeable on all aspects. I expected him to tell me that he was going to leave us to work for the TU but he surprised me by saying that he wanted to apply for the Safety Officer job at NGTE. Did he have any chance? We discussed all the issues including the major hurdle that he had to jump from Industrial to the Professional and Technical grade.

The more I thought about it, the more I was convinced that, if that step could be made, it would be a good appointment. He had the trust and respect of both management and staff but there would be a number of staff that would have difficulty accepting it. I reminded him that he would have to go through the normal board selection process against any others who would apply and I asked him if he was really serious and would not withdraw when we had to face a few emotional arguments along this route. He was always a fairly robust character and assured me that it was what he wanted and he would run the whole distance. I encouraged him to apply in the normal way. As I hoped, he went through the boarding system and proved to be the most knowledgeable and suitable candidate. He was promoted and proved to be a great success in the job.

My walkabouts seemed to help to solve quite a number of the personal and work problems in the Establishment as well as keeping me closely in touch with everyone.

The Social Life of the Establishment

There was a very active social club with all sorts of sporting activities. Its funding seemed to depend largely on the gaming machines but that was a feature of most social clubs. There were many social dances run by the various sections of the Establishment, an Old Time and Modern Sequence Dance Club, a Drama Club, a Bowls Club and where appropriate, like the Bridge club combined with RAE. We also hosted numerous events like the South Eastern Region Postal Telecomms Recreational Association Table Tennis Championships and I was invited to present the prizes at the Ryelaw Garden Flower Show.

In 1980, Rhys Probert died in post whilst Director RAE. He had been one of the great team that proposed and constructed the engine test facilities. I was appointed to succeed him in October1980. The staff at Pyestock gave me a great send-off and all, very touchingly, signed a scroll expressing their appreciation for the dedicated service during my six years as Director.

Chapter 10

Director of the Royal Aircraft Establishment.1980-1984

The Royal Aircraft Establishment was the centre of research and development into military and civil aviation and space activities in the UK and, in a rapidly changing world, played a leading role in almost all associated technologies. The S.B.A.C Show, being held there bi-annually, gave it additional worldwide publicity and was a great opportunity for additional interchanges with the scientific and military staffs from other countries visiting the show.

To be appointed its Director was a great honour, particularly when aged civil service candidates were not being viewed as entrepreneurial enough. To those who supported my appointment in the Ministry of Defence, I can only thank them. It really had always been my scientific home. I received over 90 letters of congratulation. They were all fantastic. The one that boosted me most was from one RAE staff member who said -"Your return is the best thing that has happened to RAE for a long time."

The Re-organisation of RAE.

The euphoria was not to last long! Within two days of my taking up the post, I received a letter from Sir Clifford Cornford, Chief of Defence Procurement, asking me to make an examination of the structure of RAE against a background of Government policy on Defence, the future project needs in defence, the role of the government research establishments and industry and future space and civil activities. It was a multi-dimensional task that would take some time to complete.

The study, by Lord Strathcona, of the functions of the defence R & D Establishments concluded that the emphasis of the work should move towards long term innovative and systems research, the formulation of new concepts, assistance to the military staff in the study of new concepts and the formulation of military application and the Staff Targets related to them and the assessment of their capability as systems in terms of performance and effectiveness. The Establishments should also continue to monitor extramural research, provide independent technical assessment and advice to government on UK and foreign military and civil project proposals and assistance to Systems Controllers and project managers. Effort on design, development, project support and Post Design Services activities should be reduced and contractor support for the running of major test facilities and ranges investigated. In addition, effort should be reduced in areas where the technologies were well established or where new technologies had advanced sufficiently far for the appropriate industry to take the lead in their further development for project application and production.

We were luckier than some of the other Establishments. The responsibilities for development and production of aircraft had been passed to industry in 1916, for avionics etc in the 1950s and for weapons in the 1960s.

RAE had an annual turnover of about £130m, of which £30m was repayment work from industry. It also had an extra-mural research budget of £40m for research in industry and the Universities. The manpower projections for the years up to 1984 were not, at that time, specified but, even assuming an average of the overall MoD manpower projections, it was likely that RAE, then the largest of the Establishments, would be expected to reduce from 6526 in April 80 to 5070 in April 84, something like a 22% reduction (probably optimistic). A report to HQ was required in one month!! I felt strongly that, faced with changes of this magnitude, we had to select areas where the work should be retained and perhaps enhanced, select new areas for study and opt out of other areas completely. In addition, it was important to consider the skills mix required and not just the numbers of staff alone. Finally, the proposals would only be acceptable if the broad balance of the grade structure was maintained with perhaps greater than average savings at the senior management level.

Two weeks of meetings with my Management Board showed that, if we were to make recommendations in line with Strathcona, we needed more time, firstly to make recommendations to HQ and then to get agreement both at HQ and locally. In addition, as part of the Rayner/Strathcona activities other studies were in progress. Harold Robinson, DD(E), who had been acting Director since Rhys Probert's untimely death, told me that he had been studying the future of the P.E. Airfields -Farnborough, Bedford and Boscombe Down. He had the draft report ready to publish that recommended the closure of Farnborough Airfield. I was very familiar with the arguments, having done the airfield study some 8 years earlier. I suggested that he should not show it to me and that he might take a month or two before publication i.e. as slowly as possible, to give me time to tackle the other problems first. Eric Rogers, DD(A), was starting a study on the contractorisation of the Wind Tunnels (200 to 350 posts). A similar exercise with NGTE and Rolls Royce had led to nothing. In addition, we were required to examine the possibilities of contract operation of the Concorde Structural Test Facility and putting more of the aircraft installation and maintenance work out to industry. Further contractor operation of the ranges was also to be studied.

In March 1981, after negotiating an agreement with CERN, we issued a report- Proposals for the Future Role and Organisation of the Royal Aircraft Establishment. This report was sufficient to get the re-organisation activities started but there was a great deal of work to be done to keep the moral of the staff as high as possible, keep both the important technologies in being and get new technologies started with sufficient effort and funding devoted to them to ensure that their output would be timely. At the same time, we had to avoid industrial relations problems during the period of considerable reorganisation and contraction. I hoped that it could be achieved by briefings to the staff at every stage and seeking their views on the proposals and their suggestions for advantageous changes that needed to be considered.

The changes involved the split of Flight Systems into two Departments, phasing out some of the work of Engineering Physics Department and re-locating that part to be retained, the combination of Materials and Structures Departments into one department and the formation of a New Concepts Division in Space Department to initiate new activities that could be undertaken anywhere in the Establishment where ever appropriate facilities were available. As the reorganisation progressed, new items were added including, (as will be described later) the integration of NGTE into RAE, and the transfer of Trials Guided Weapons (TGW) from HQ to Farnborough.

All this change was combined with the gradual reduction of senior posts as people retired, of one Under Secretary post**, one CSO (B) post 1 1/2 *, three DCSO posts *, and 9 SPSO posts with overall numbers targets reducing year by year of 10%, 5%, 5% and 5%, i.e.25% in 4 years. I thought that it was a mammoth task that was going to need the support of everyone in the Establishment.

It was arranged that, on one morning, I would brief all Heads of Departments, Superintendents, and their equivalent managers in other areas. They would leave that meeting with a copy of the report. It would be their job to brief their staffs, whilst I briefed the Staff Side and Trade Union Representatives. In this way, it was hoped that everyone got a complete picture of the changes proposed. It was also released to the press. As expected, there were national and local newspaper reports of fears of RAE redundancies and the introduction of contractors.

By August 1981, the TU defence wide, were protesting against the probable staff cuts. The RAE and NGTE Staff and T.U. representatives were very balanced in their comments. Follow-up meetings were held with the senior managers and the staff and union representatives to hear the reactions to the plans and to answer their questions as far as we could. It was a multi-aspect plan that could only clarify as each part of it was brought to completion.

In the conclusions of the document on the reorganisation proposals, said, "The proposals for the programme of reorganisation set out in this document involves a very heavy management workload and, if agreed, will take time to implement. Much of the detail has yet to be examined and, in particular, the implications of the organisational changes in the Research Departments have not yet been carried through to the Chief Engineer's and Secretary's Departments and will need to be studied. These proposals are being put to the staff of RAE, the Aviation Industry, Systems Controllers and other interested parties. All comments will be welcome as an aid to the various decisions that have to be taken and should be addressed to D/RAE." The staff made numerous suggestions, particularly on where they should to be relocated. As a result the plan was refined and improved. Scientifically we were stronger as a result of the reorganisation and the programme much more directed to our tasks and the future defence requirements.

Below the surface, there were many other problems to which solutions were not available. The staff numbers were to be reduced but there were no mechanisms other than wastage and wastage and bans on recruiting by which to achieve them. Both options were anathema to the plan we had produced. For over three years, we had wastage, bans on recruiting, and surplus lists. It was not until late 1983-early1984 that MoD introduced some early retirement schemes and we were then able to recruit sufficient graduates to sustain the high technology programme.

Another *minor* problem was the continued practice of the allocation of staff numbers and funding separately. We were under-funded to do the work with the staff in post. I had to hope that I could depend upon AWRE to under-spend and that my over-spend would be absorbed in the CERN financial reconciliation in London. AWRE never let me down!

The Airfield

I was sure that Harold Robinson, given his terms of reference, had reached the correct conclusions. It was: - " The closure of any of the airfields would take several years but that the closure of Farnborough was the option that showed least disadvantage, although it involved severe problems and that, in the light of this study, Boscombe Down and Bedford airfields should continue as at present."

The airfield at Farnborough was seriously under-utilised and military flying would be decreasing. It was just impossible for me to envisage the origin and centre of aviation and space research in the U.K. without an airfield. Also where was the SBAC Show to go? I knew that there were many financial managers with their eyes on the whole of the RAE site that they saw as ripe for development. It was almost possible to hear the money clattering in. I was sure that it was a mirage. The land had been common land and was commandeered for military purposes in WW1. Although the opportunity to reclaim the land for the community had long passed, there were legal precedents that indicated that if a proposal was made to sell the airfield and develop the land, there was a high probability that it could be successfully resisted. In addition, the Rushmoor council would have views on the danger that such a development might overload the local infrastructure.

The possibility that the airfield was under threat was widely known. By the summer of 1981, protest marches were organised. Locally, a public meeting was called on 27th July 1981, in the Farnborough Town Hall for people to air their interests, support and/or fears. A property developer with a keen interest in aviation, Alan Curtis put forward his plans, using the assets of the Shell Pension Fund and the huge Slough Estates property firm, to develop part of or the entire 1400-acre site and add substantially to the jobs there. Dan-Air Engineering was also interested in using the airfield and the facilities. MoD was fairly tight lipped about any proposals, saying that they would make an announcement when they had made their decisions. They would neither confirm nor deny that the closure of RAE was being discussed. Julian Critchley, the local M.P. attended in order to report back on the discussions to his Minister.

In London, my desire to keep the airfield open was met with considerable scepticism. But I managed to persuade Sir Geoffrey Pattie, Minister of Defence for Procurement and Sir John Charnley to agree that the future options needed to be examined using much broader assumptions. We, of course, were delighted. To get the proposals together, many people would have to be consulted and there was considerable work to be done.

By October 1981, I had agreement "to examine by what means and to what extent Farnborough airfield could be opened to General Aviation use." By November, we were able to publish a paper, " The Introduction of General Aviation to Farnborough Airfield". Two options were proposed.

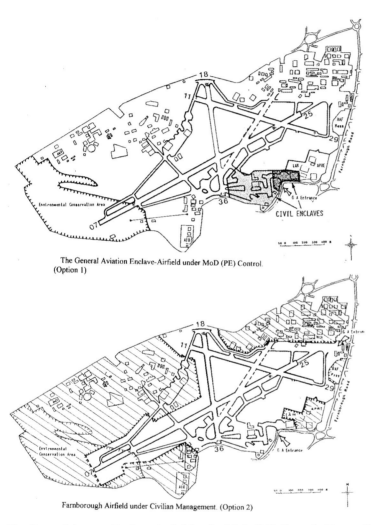

The General Aviation Enclave-Airfield under MoD (PE) Control. (Option 1)

Farnborough Airfield under Civilian Management. (Option 2)

In the first, a fifty-acre site on the south side of the airfield was identified, some three-quarters of which would be for business aviation and one quarter for light industry. RAE would remain in control of the airfield and the G.A. aircraft would enter via sliding gates.

In the second, airfield and civil enclave were one entity. RAE would not be in control of the airfield and would then pay fees for its use.

It was estimated that the number of movements at Farnborough, which had fallen to about twenty thousand per year would, with the influx of 40 General Aviation aircraft, increase again to the levels of the mid-1960s. However, calculations of the Noise Number Indicator indicated that the noise impact arising from these additional G.A. aircraft movements would be negligible outside the airfield boundary and inside the airfield boundary the change would be very small.

By October 82, we had published a paper on the Proposals for Future SBAC Arrangements at Farnborough, and, by January 1983, a paper on the Possible Move of CAA Offices to Farnborough. (I think that it is important to say here what a considerable contribution many of the Farnborough staff made to this work, particularly Ken Turner of the Management Services Unit.)

There were three other areas for development. Firstly, if the radio aerials and equipment could be transferred to the western end of the airfield, the site of the Cove Radio Station could be sold off for housing development for a sizable sum. Secondly, the Bramshot site that had become surplus to requirements could be sold. Thirdly, the SBAC site could be developed into a more permanent exhibition site accommodating SBAC and other exhibitions. Representatives of SBAC visited the site several times to look at particular issues of concern. They were; - the adequacy of parking on public days, the detail of the funding, whether SBAC would be indemnified if their income from the 1984 show fell below the receipts which they would require to break even, whether they could lease some of the enclave or perhaps take part in the running of the general aviation area, the ideas on developing the permanent exhibition site, etc.

There were a number of other difficult issues to resolve. MoD would not guarantee the availability of Farnborough to SBAC except on a two-year basis. Without some security of tenure, SBAC were reluctant to commit funds to develop Farnborough and were even more reluctant to put up some permanent exhibition halls and become responsible for running and funding an exhibition site. The Army would not hear of moving their exhibition, from their site about three miles down the road, to Farnborough. (It was not until a few years later, when there was a financial squeeze on exhibitions that this proposal was being taken seriously.)

Between November 1981 and March 1982, I made presentations on the airfield options to two County Councils and five local Councils. The presentations were well received. They involved some amusing incidents. I presented our proposals to the Rushmoor Council and then again at a public meeting attended by some 300 members of the public in the Technical College at Farnborough. By that time, Residents Against Increased Noise (RAIN) had become an established, but not very effective, protest group and wanted to make their presence felt. In addition, the Russian Assistant Air Attaché, Colonel Gennadiy Primakov attended and commented, "As my country's Air Attaché, I am interested in aviation. Farnborough is a well-known international place. It is important for me to see what will happen to Farnborough if flying stops. There is talk of it closing. What would happen to the air exhibition?"

At Fleet, the presentation was made on January 7th 1982. Just as I started the presentation, the orchestra of the Christmas Pantomime performing on the floor below, struck up for the opening overture. There was no alternative but to shout my presentation to the audience of some 200. In closing my remarks, I said "It is emphasised that this presentation gives examples of typical activities for the future which have been derived from RAE's own studies and from informal discussions with interested parties. They will also, of course, be influenced by our consultations with local authorities. There is a prospect that General Aviation could be introduced to Farnborough giving advantages of increased efficiency in the use of the airfield and its facilities and increase job opportunities in the area with only a small degradation to the current environment. At this stage consultations are in hand which will enable future plans to be drawn up with the confidence that progress can be made without serious diversity of views." How optimistic I was!!

At the end of the presentation, the first questioner got up and made a five-minute speech opposing the proposals he finished by saying that he had never heard so much rubbish in all his life. Until that point, I had no clear idea how to answer his assertions properly and effectively. That last phrase gave me the opportunity to start my answer by saying that I was delighted that he could hear me at the back. The audience burst into peals of laughter and he left. From then on, there was a very useful period of questions and answers.

The views of the local authorities were, on the whole, favourable but they required numerous safeguards. Nevertheless we had sufficient to proceed with the next stage of the negotiations. It was the unanimous view of all authorities that RAE must remain responsible for the airfield i.e. Option 1.

We had to establish that there was a real interest in developing the site and we had to continue to attract that interest whilst trying to establish the mechanisms whereby the SBAC show could continue to occupy the airfield for about two weeks every second year.

The papers were distributed quite widely both inside and outside the Establishment so that everybody was well informed of the progress. Negotiations continued very slowly. The General Aviation developers, who were interested wanted, a twenty year lease to justify the capital investment and SBAC wanted some security of lease to enable them to justify more investment in the show area. Eventually, MoD agreed that a 20-year lease would be granted and SBAC would have security of tenure for the show until 2002. I thought that Rushmoor Council, the most affected local authority, did a superb job in balancing the conflicting interests whilst at the same time doing their best to ensure the continuing prosperity of the district.

In order to try to get a balanced argument on noise, we invited residents under or near to the flight path to have noise meters installed in their gardens. We had many offers and managed to get instruments for most of them. We then hired some representative business aircraft to fly circuits and landings so that the noise could not only be heard but also measured. The results supported our predicted noise contours.

After a very heated debate, Rushmoor Council finally recommended that: -

i) MoD's proposals for general aviation at Farnborough airfield be approved,

ii) MoD should be told that the Council supports the introduction of light industrial use within the civil enclave extending to 22 acres,

iii) the council's planning officer should prepare a planning brief on the civil enclave at the airfield for discussion at the next planning committee meeting,

iv) the council informs the MoD that it requires an undertaking that there will be no increase in the numbers and weights of aircraft, movements and hours of operation.

Within RAE, Dave Cook, the Chief Engineer, found enough money to fence off the enclave so that the many visitors to the site could be handled without having to pass through our security system. Because the Councils were so keen on RAE continuing to control the airfield, sliding security gates, to allow G.A. aircraft access from the enclave to the airfield, were installed.

Despite many ups and downs since those early days, the proposals have made continued progress. Many people have devoted great effort to its success, including the public inquiry in 1997 held in the Rushmoor Council offices. It was a great pleasure to see that, in 2002, TAG, a leading provider of business aviation services, has leased the airfield and has approval to develop it by building a new terminal, new hangers and a new control tower. Aviation will stay at Farnborough and continue to enhance the reputation and the prosperity of the area.

Contractor Operation of the Ranges.

The RAE range sites were at Aberporth, Llanbedr, West Freugh, and Larkhill. The task was a very specialist one, using highly devoted teams who had an excellent record of achievement. Over 90% of the work was of a development, acceptance or service trial nature for land, sea and air weapons. The main considerations were: - the facilities needed up dating, their location was in areas of relatively high unemployment and limited job opportunity often leading to rather distorted manpower skill distribution. There was nothing in the range operation that made it essential to retain as an integral part of RAE but there was considerable benefit in having them associated with a large establishment. Thus they could become part of AAEE who had responsibility for acceptance and release of aircraft and associated weapon systems or they could become part of a wider Ranges Authority embracing MoD (PE) Ranges. Whatever the decision, RAE could continue to provide appropriate advice and support as necessary.

Irrespective of the ultimate responsibility for ranges, there appeared to be advantages in eliminating the overlap between the Director of Trials and Guided Weapons, in London and RAE ranges. DTGW, under Air Cdre "Fitz" Fitzpatrick, was responsible for supplying the targets for the ranges including, Jindivicks, Stilletto, the supersonic target, droned Meteors and Sea Vixens and the towed (on 30,000ft of cable) Rushden Targets including the testing of the sea-skimming version. The range staff had to bring together all the programme management issues including telemetry, flight termination systems, scoring systems, transponders, flight data systems as well as the major procurement tasks of the radars and the new Break Up Predicted Impact Area System.

We, therefore, proposed that the TGW responsibilities be transferred to the RAE with the loss of one DCSO post, that the contractor operation of Llanbedr be expanded, that West Freugh be contractor operated with only ten RAE staff remaining as the management team. Aberporth would continue to be staffed by RAE because of the commercial sensitivity of some of the work but with more contractor help.

The most difficult of those tasks was the introduction of the contractor operation of West Freugh. There was about 110 staff. RAE would retain about 10 to manage the trials programme. In their bids for the work, the contractors were encouraged to indicate to which of the RAE staff they would offer jobs. It appeared that they wanted to retain the bulk of the work force because they held the skills that would make the future operation of the range successful. That amounted to about 70 staff. Many of these staff would be better paid as it was possible for the contractor to combine jobs, such as camera men and drivers, e.g. they could drive themselves to the remote camera sites and operate them. In this way, posts were saved and responsibilities expanded. Some ten of the staff wished to retire and ten wanted to seek posts at other locations.

That left 10 staff with a problem. We set up an operation to help them find new work but it was not easy. The T&GWU proved to be particularly truculent. One day someone rang me to tell me that if I continued with the contractorisation programme, they would bring to bear hostile actions that I had not experienced before. Sensible discussion was impossible. At the end of the conversation he said, " I'm no averse to a bit of brigandry!!" I replied that it was lucky that I had taped the whole conversation. The phone went dead. From that point on the operation went smoothly until the day of the handover when one of the small number of current contract staff, due to be replaced, destroyed a number of important operating tapes. It had nothing to do with the T&GWU.

The Contractor Operation of the Wind Tunnels

Where large facilities, e.g. wind tunnels, were in support of one major contractor, the Strathcona policy was that they should be part of the contractor operation. MoD(PE) had tried this with Rolls Royce and the Pyestock Engine Test Facilities during my time at Pyestock and had failed to reach any accommodation with the company other than increasing the charging for engine testing. The wind tunnels were in a similar relationship with British Aerospace. Eric Rogers DD (A) was already studying with BAe and other interested parties which of the wind tunnels were needed to maintain our national capability, taking into account those available in Europe and in the USA. They identified what was needed and its location, whether it be in the company, in the RAE, or in an independent such as the ARA at Bedford, or an overseas facility. This requirement list was agreed.

There were two alternatives to be considered for contractor operation: -

a) the whole site and tunnel complex could be handed over to the contractor, who would be `responsible for producing and perhaps analysing the test results according to an agreed programme or

b) the tunnels remained an RAE responsibility but with contractors operating the tunnels and providing engineering and other support. A small number of RAE staff would remain to evolve the tunnel programmes, supervise the tests and evaluate the results.

Our view was that, at Farnborough, the wind tunnels were so interlaced with the other operations on site that contractor operation would be complex and may not even be efficient. Whereas, at Bedford, the wind tunnel site was self-contained and the outcome, whatever it was, would be clearly visible. We decided to tackle Bedford first. The next task was to establish the manning level required for future operations. BAe assisted with this task and there was considerable agreement between us. By early 1984, John Stamper, a Director of BAe and an old friend of mine, was with me at Bedford reviewing the work that had been done. He said" Well, what do we do next? Get a team to come in and value the site?" My favourite "imp" then took over the conversation and I said, " I don't think we need to do that John. You can have the lot for £10." I guess that I was as tired, as was the rest of the staff at Bedford, of the uncertainty that these exercises caused. He agreed to take the proposition away and consider it. Some three weeks later he rang me to tell me that he had briefed the BAe Board on the whole exercise and they had decided to turn down my proposition. Although he never gave the reasons, I have no doubt that they had looked at the running costs of more than £5m/year and decided, as R.R. had done, that as this cost would largely come straight off the bottom line of their accounts, the longer they avoided this cost the better.

The Concorde Structural Test Facility.

This facility, built to establish the safe fatigue life of the Concorde, was manned partially by BAe staff responsible for the test airframe and partially by RAE staff responsible for the plant. The Strathcona Study recommended that contractors operate it. By 1982, the British and French Governments had decided that no more public funds would be invested in Concorde. Any further support work for the aircraft would have to be funded by British Airways and Air France. Their study took several months. Eventually the airlines had decided that work in the Concorde Test Facility was to be discontinued. By that time, enough testing had been done to clear the aircraft for airline service, at the then current utilisation, until 2002 i.e. nearly twenty years. (Today in 2002, it is cleared for operations to at least 2010 as the comparisons between the test specimen results and the in-service aircraft usage and experience has been good enough to reduce the 3 to 1 safety margin of the original test programme and thereby extend the operational life of the aircraft.)

The closure of the facility was the first real opportunity that we had to declare a programme of voluntary redundancy. I decided that we should identify the 38 posts that would be made surplus by the closure of the facility, list those who were currently filling them and if they did not want to be redundant, re-employ them elsewhere. The redundancy slots would then be offered to others in the Establishment to volunteer to be redundant in their place. I talked, in confidence, to the Whitley Council representatives three working days ahead of making the announcement. It gave them time to consider the plan and ask questions if they wished. There were no questions and the redundancy was announced.

One day I wanted to visit a Department on the other side of the airfield. We had a system of radio cars available for everyone to use internally to save time and, for safety, avoid employee cars on the airfield perimeter. On route the lady driver asked me if I was going to make her redundant. I asked her whether she had volunteered. She said that she had. I asked her how old she was and she said sixty-two. I replied that I could not promise but it was quite probable. She said "Thank you Sir." The whole process went smoothly and eventually 78 volunteers left.

Chief Engineer asked me how much of the facility he should put into "care and maintenance" and how good this needed to be in terms of the time required to recover to full working condition should further testing be required. I told him that it should not be preserved in any way, as we had no money for it.

The Integration of NGTE into RAE

Wason Turner, who succeeded me at NGTE and did an excellent job there, was due to retire in April 1983. CERN asked me to integrate NGTE into RAE. There was the immediate prospect of saving of one ** post, with subsequent savings in support services such as administration, fire and medical services, apprentice school etc. The people at NGTE knew me well and I had an in-depth understanding of their strengths and problems. The research departments went under DD(A) and the facilities under Chief Engineer. To preserve the history of the Establishment, I had an RAE Crest put on the gates by the side of the NGTE crest. I think that it pleased everyone.

When I left NGTE, there had been an ongoing study under Colin Hockinholl on a new bonus system. It had been brought to a successful agreement. In addition, as a result of the H.Q. negotiations with R.R., a new charging system for engine testing was in place.

An outstanding difficulty was the integration of the Staff Unions and Industrial Unions Whitley Councils so that meetings with management could be handled in the normal way. Unfortunately, the arrangements could not be agreed. Eventually in some despair, I wrote a paper on how they should organise themselves. I was about to hand it to them when I remembered that perhaps I should let the Official Side of Industrial Relations in London see it first. I sent it to HQ with a note explaining the situation and said that if I did not hear from them in two weeks I would assume their approval. On the 14th day I had heard nothing and issued instructions that the papers were to be issued at the start of the 15th day. At 10.30am that morning, they rang from London to tell me that they did not agree with my issuing the paper, as it was no part of management's task to tell the Staff and TU sides how to operate. I said, " Pity it has been issued." I am happy to say that the paper broke the deadlock and, within a month, we were operating normally.

It did not solve all the problems as I wanted to integrate the fire services into a new designed building on the north side of the airfield so that they could provide the fire and rescue service equally well to both sites. Unfortunately, it was an issue that I had not solved by the time I retired.

Before leaving the NGTE story, I must recount a little incident that caused me much amusement because it happened off my patch. Known only to a few, we had always helped with the training of the Army by allowing them to practice attacks on us. It was not only a test of our security systems but it gave them practice in planning and carrying through their own operations. On one occasion we used Bramshot. We had detected their intrusion one night and decided to install alarm systems in some of the buildings the next day. The Army, watching, from their hides close to the site, thought that what they saw were normal activities and got a considerable shock on the next night when, during their intrusions, the alarms went off.

Wason Turner must have agreed to a similar exercise at Pyestock. On December 9th 1982, the Star headlined "Storm Troopers brew up tea break storm". The gun-toting soldiers, who landed by helicopter in a night raid on a local defence establishment, burst into the wrong room, terrifying the workers having a quiet cup of tea, were from the Army, according to the allegations of an angry worker. According to one report a hooded and armed man burst into the rest room tipping a man out of a chair and brandishing a revolver. Officially it was a very low-key incident and Wason is quoted as saying, quite rightly, "Least said soonest mended".

Staff Inspections etc.

To keep the Comet Flying Laboratory in the air required the usual supply of spares. They were becoming difficult to obtain, as we were almost the last operators of the aircraft. When Dan Air disposed of their Comet fleet, we were able to buy their spares at a very low price, about 10% of the usual cost. The MoD inspectors had wanted to come and inspect RAE for some time. In the midst of this massive reorganisation, they were the last people I wanted to see. With some reluctance, I agreed that they could come and inspect the stores, provided that they report their findings to me first. That requirement did not go down well. During that inspection they found that the Comet stores were held, as they had been delivered, in the crates. Each crate was marked with the particular stores that it contained. This arrangement was not standard practice and their report contained some heavy criticism of these ad hoc arrangements. They recommended that the spares be stored on shelves in the normal way. I pointed out to them that buying the shelves and making these arrangements would cost more than the stores had cost.

They pressed for further inspections of the staff in the Research Departments. I kept refusing. Eventually, I decide that PUS would hear of the problem and I arranged to go and see him and invite him to Farnborough so that he was briefed on all that was happening. He came and was very surprised by the changes that were in train. He so thoroughly enjoyed himself on the visit that he came back again. He wrote me a very encouraging letter thanking us for the fascinating day and adding,-RAE's scientific work is plainly as good as ever; and, if I may say so, you are tackling the management problems that the whole Department faces with a measure of ingenuity and flexibility that I do not find everywhere. On a subsequent visit, to Bedford, he flew the flight simulator. I never had any more requests for inspections.

By the end of 1983, the new organisation had settled down and the RAE locations were still much as before. The RAE Management Board, at the end of 1983, was photographed (below) in front of the cast aluminium replica of Cody's Tree upon which birds never rested, probably because it felt strange to their feet. Left to right-front row, Dr.Derek Dawton DD (W), myself, Peter Whicher, DD (E) and the rear row, Dave Cook Chief Engineer, Group Captain Charles Officer Commanding Experimental Flying and Dennis Blazey Secretary.

The RAE Locations

The RAE Board of Management.1984.

Somehow, Eric Rogers, DD (A), missed the occasion. His photograph is inset at the top of the picture. After all of these dramatic changes to the Establishment, I was convinced that it was scientifically and organisationally stronger and better able to operate in the future aerospace environment in defence and civil aerospace. I felt justified in spending so much of my time to achieve this objective.

THE TECHNOLOGY

The technology was exciting but it would take several books to do it justice. What thrilled me about the reorganisation was that, despite the cut in resources, we preserved and in some areas enhanced our most forward-looking capability. I will describe only those areas where I was most involved or had the greatest interest.

Computational Fluid Dynamics. CFD.

Since the early 1950's, Aerodynamics Department had been developing the mathematical techniques for calculating the flow over wings and bodies. The demands of Concorde emphasised the importance of the work and CFD gave, not only the means of creating highly efficient aerodynamic designs, but also the potential for great cost saving by building fewer tunnel models and reducing the expensive wind-tunnel testing time. The technique required a new technology supercomputer. At the time, the only one available was the CRAY at AWRE. We were able to hire time on it but it was inconvenient and involved significant security problems. We had plans for a new computer complex and had been in discussion with the Property Servicing Agency for some time. Unfortunately funds for new building were being cut on a regular basis. Repeated failure to start this project was incredibly frustrating for all.

Eventually, it was clear that we had the funding for the computer, we had the building approved and we were ready to start. At that point, PSA rang to say that the funds had been cut and a start was not possible on the building. Examination of the problem showed that, if we deleted the offices, we had enough money to go ahead. From then on the progress was good and we soon had our own facility containing a CRAY 1S together with a Honeywell Multics system. The capability has been so valuable that the contract for building the airbus wings was won on the foundation work done in this and earlier facilities. In addition, the new building contained some special computers devoted to computer graphics, MODAS flight test data replay facilities and equipment for handling administrative data.

There was one amusing incident. When it became common knowledge that the building would not have any offices, I was asked about it at one of the regular meeting of the Whitley Committee. I described the events relating to the computer building and why it had been built without offices. They then asked me to explain my policy on buildings in RAE. Many of the buildings in RAE were old, inefficient, and often only partially occupied. As they were unsuitable and costly, I really wanted to get rid of many of them. Therefore, my policy was to call in the Chief Engineer every month, give him a shot of adrenalin and tell him to go and knock down another building. When we had all stopped laughing, the question was not raised again.

Flight Systems

Normal white light view of Tower Bridge.

The same scene as it appears in the infrared waveband.

Daylight Photography and FLIR Pictures at Night.

Thermal Imaging had been a key part of our work and the use of Low Light TV, Night Vision Goggles, and Forward Looking Infra-Red (FLIR) had been going on for a number of years. It used the technology developed by RSRE and extended its use from reconnaissance to its use in helicopters and fixed wing aircraft flying at low level at night and in bad weather. FLIR works by detecting very small differences of temperature (< 0.1deg C) rather than reflected light. It can operate in total darkness. Because the wavelength of operation is 20 times longer than that of light, it can see through smokes, haze and fog. The pictures, taken from the Varsity equipped as a flying laboratory can be interpreted as easily as daylight pictures. There was considerable interest in the work- everyone wanted to fly the aircraft fitted with the complete system. It was used for demonstrations on two nights a week.

When pilots turn the aircraft, they look ahead of the turn to see the terrain and any other hazards in their intended new flight direction. Night Vision Goggles (NVG), fitted to the helmet gave complete freedom of head movement. However, there was an incompatibility between the NVG's and the cockpit lighting that caused the NVG's to over-glow and loose vision. The new Aviators Night Vision Imaging System (ANVIS) goggles, manufactured by Bell and Howell, were a considerable improvement on the earlier models. They were neater and lighter and sat some 15 mm in front of the pilot's eyes whilst still presenting a full 40-degree field of view

Mick Law of Flight Systems, Farnborough, Helicopter Section wearing the night vision goggles.

The ANVIS Low Light Night Vision Goggles.

through the intensifier tubes. This meant that the pilot could look through the goggles at the intensified image of the outside world and round them to read his flight instruments and maps.

The only modification required on the aircraft to make the system fully compatible was to change the cockpit lighting system. The night goggle intensifiers were particularly sensitive to red, and near infrared, radiation and it was at this end of the spectrum that the filament cockpit lamps radiated most efficiently. Dimming the cockpit lights merely compounded the problem. The high levels of near infrared reflecting round the cockpit prevented the pilot from seeing the outside world clearly. The solution devised by the helicopter section was to illuminate the cockpit and instruments with visible light from which the red and near infrared had been filtered out. This change was achieved by turning off all existing instrument lights and flooding the panel with light passing through specially dyed and optically coated filter glass giving a blue-green colour. The change resulted in pleasant cockpit lighting whilst still retaining the ability to recognise colour on the instruments and on maps. This combination avoided the NVG bright-ups and the pilot could still read the instruments by glancing round the eyepieces. The combination was a considerable success.

With the two-seat Buccaneer fitted with the latest equipment we were able to demonstrate the ability to fly at low level over flat and mountainous terrain at night and in poor visibility. It was demonstrated to many senior RAF officers and it never failed to impress them. On one occasion we had a visit from the Deputy Commander of the USAF 2nd Tactical Air Force. On the night that he was due to fly, it was very foggy. He fully expected the sortie to be cancelled but our pilot assured him that it would go ahead. After take off, the pilot called RAF Odiham, closed due to fog, and asks permission to make a low fast run down the runway. Odiham agreed and asked if he wanted the runway lights switched on. He replied "No thanks". He completed his run. They heard him but did not see him. The US General was very impressed. At that time, they had nothing like that capability in the USA.

RED OWL (Remote Eyes in the Dark Operating Without Light) enabled a pilot to

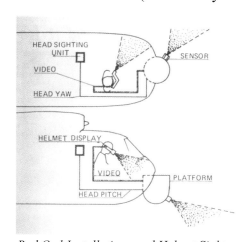

fly a helicopter at very low height, at light levels less than clear starlight, and undertake search and rescue operations. The display was mounted on the helmet of the pilot. A low light level TV or infrared sensor installed below the nose of the helicopter supplied the image to the pilot. The novelty of the system was the way this sensor follows the head movement of the pilot, providing him with an uninterrupted wide-angle vision of the world outside. He could effectively see "through the floor" of the helicopter as his field of vision extended from 100 degrees to left or right of centre and from +20 to

Red Owl Installations and Helmet Sight.

-100 degrees in the vertical plane. The pilot's head angles in pitch and yaw were fed to servos in the sensor platform so that the camera accurately follows the pilot's sight line. At its very heart was a fast acting servo, swivelling at over 1000 degrees per second, essential if the pilot was to avoid becoming severely disorientated by any lag in the response of the electronic eye. In the infrared mode, the system was able to detect and display objects having temperature differences, relative to their surroundings, as low as

0.1 degrees C. It meant that on a pitch-black night over the sea, the pilot could detect someone in the water and a rescue could be made.

The system was displayed at the SBAC show in 1982. It involved taking the visitors into a dark room so that the displays could be viewed correctly. Unfortunately, at SBAC shows, there are numerous attempts to steal items of equipment and we had to take security measures to protect the displays. In this instance, an IR camera was installed in the dark room. In the event nothing was stolen but some of the activities of the visitors, particularly couples, were quite unexpected.

Civil Aviation Navigation and Colour Cockpit Displays

The BAC1-11 programme at Bedford was 90% funded by Department of Industry and 10% by MoD. Everyone involved with the programme recognised the urgent need for successful development and sales for the continued health of the UK Avionics Industry. Over the years this programme had involved: -

-making better use of existing navigation aids and, by the increased capability of airborne computing.
-investigating techniques for approach and landing in visibilities as low as 150 m
- changing the displays from monochrome to colour.

A combined RAE/industry team took the aircraft to the USA, the biggest avionics market in the world. We had thought that we were beginners in the field but found that our close knit, multi-capability team was a world leader in this integrated technology. The demonstration of accurate navigation in 4-dimentional space (including time), involving errors of less than 1/4 mile standard deviation over the whole route with a target time slot error of about 10 seconds, was unmatched.

Fly-by-Wire and Flight Simulation.

The RAE programme on "fly-by-wire" techniques for flying the aircraft had been progressing for a number of years. It started with a two-seat Hunter allowing manual control by a safety pilot and progressed to the Jaguar carrying four computers, which had to have very high reliability since the system was flight critical. The potential gains to civil aircraft by adoption of similar systems were substantial-for example, the reduction of fatigue loads in rough air, promised much improved life cycle costs through the impact on airworthiness.

Since I first flew the Link trainer in1941, I have found flight simulation a fascinating subject not only for training without risk but also for the investigation of the man-machine interface. The use of multiple computers presented the opportunity to explore and apply new control laws that would make the task of the pilot both safer and easier and by the use of active control technology operate an unstable aircraft that the pilot would be able to fly.

Peter Whicher DD (E) was pressing for new programmes on two fronts. Firstly to get a new motion simulator at Bedford and to get approval for a new modified aircraft control system, Vector Thrust Aircraft Advanced Flight Control, (VAAC), established as a technology and as an experimental system installed in a Harrier. I was so enthusiastic about the potential of these new systems that every time I went to Bedford, I tried to find time to fly the simulator set up to study the pilot problems of using the proposed system on the Harrier.

In the standard Harrier, in normal forward flight, two controls are essential, i.e. the stick and the throttle to control the speed and the height. In the hover, three controls are required, i.e. the stick, the throttle and the nozzle angle. In addition, any change made to any one of them has an interaction on the other two. The pilot has only two hands. Nevertheless, when pilots can concentrate entirely on the task, i.e. at SBAC show, they demonstrate the aircraft beautifully. However, if he is trying to land in poor visibility, in a storm, on the pitching deck of a carrier, the workload is dramatically increased and the pilot is unable to cope. The VAAC system was intended to overcome that problem.

In normal flight above 130 kts, the VAAC flying controls act in the way of a normal aircraft.

In hovering flight below 60kts, the stick now controls the height demand (and not the pitch angle of the fuselage which can, if required, be changed by a trim switch on the stick) and gives a rate of climb or descent. The throttle gives forward (or rearwards) velocity. There would normally be two indents on the throttle for the pilot to feel, one giving zero air speed and the other zero ground speed i.e. hover.

These two flight control modes are blended between 60 and 130 knots. Although in the hover, the functions of the cockpit controls have reverse responses to the standard Harrier, the transition between the two flying modes was smooth and natural. The funding was agreed for the installation in the Harrier for thorough flight-testing and comparison with the simulator results.

It was wonderful to see that the VAAC Harrier flew successfully in December 1991 to the great credit of all concerned.

Space Department

From the flight of Sputnik in 1959, the potential for the use of remote sensing from space has been exploited. There were many spy satellites used to assess both the enemy economic and military strength and watch the development of future threats. The civil use of space for surface observation (and sub-surface observation) grew at a fantastic rate. In 1980, the National Remote Sensing Centre (NRSC) was formed at Farnborough to foster co-operation between the earth observation capability of the satellites, in both the passive and active modes, and the users. It had applications in the study of land use, coastal processes, oceanography and meteorology. Combining these capabilities with the multiplicity of sensing techniques and spectra and the very rapid developments in image processing presented the possibility that subjects like mineral resources, agriculture and forestry, mapping (even depths below the surface of the sea) could be studied. In 1983, the Department of Industry predicted that, by 1990, the remote sensing industry would have a turnover of at least £250m a year.

The use of space satellites for remote sensing seemed to be limited only by the financial support available. The amount of data that was produced in a single satellite pass was enormous. Add the multi-spectral capability of some satellites, in particular, the synthetic aperture radars, rapid data processing was one of the most important requirements. The radar satellites were able to detect, not only ocean waves, but currents, oil slicks etc. and features at the bottom of the sea, such as reefs and sandbanks, by the disturbance they create on the surface. The images could be obtained independent of cloud cover, a characteristic that was important in some applications. RAE became the National Point of Contact (for the exchange of remote sensing information) with the European Space Agency and was a member station of Earthnet, a European network of ground stations for the reception of data from satellites.

It requires another book or two to describe all the technology that was being developed in the various departments. I can only mention them but they all have their place in history to the great credit of those who devoted themselves to them.

A few examples are: -

Materials and Structures Department had made a major contribution in the carbon fibre and composite materials field industry including Courtaulds and Morganite, and had provided a high level of technical support for industry. The UK industry was slower to develop its markets and production capability than some of our overseas competitors, notably, the Japanese. In 1981, the USA produced 330 metric tonnes and the Japanese 700 metric tonnes whereas Courtaulds produced only 120 metric tonnes and sold most of its products into the non-aerospace market. That was a great disappointment to us. However, we went on to develop battle damage repair systems, for wings, tails and fins, which could be used in the field.

M & S Department led the field in the development of lithium-aluminium products, again in close collaboration with UK aluminium industry. This was much more successful nationally, with RAE doing the research and the industry developing viable product routes. The USA was sceptical and jealous of these advances. At one stage, they claimed that in an intense fire, the metal would burn. Tests were made in which near molten ingots of lithium-aluminium were dropped into water to demonstrate that they would not spontaneously burst into flame. The tests demonstrated the safety of the product and produced some wonderful metal sculptures.

The Weapons Departments were each contributing in a major way to the advances in technology that would enable us to develop more accurate, versatile and reliable weapons. The development of low cost, reliable and accurate Terminally Guided Sub Munitions was a high priority at the time.

In the nuclear area, AWE was responsible for the nuclear package and RAE had always been responsible for the rest including the performance of the delivery vehicle, the arming and fusing etc. Chevaline, the UK independent nuclear deterrent was a vitally important programme. Being R & D authority for the bus and the re-entry vehicles tested our capability to the limit. Nevertheless, it turned out to be a very successful programme.

The Falklands

The crisis started from the appointment of General Galtieri on Dec 22nd 1981 when he claimed considerable concessions in the Falklands. The indigenous population wanted no concessions to be made. When no progress was made, Argentina considered itself free to choose a procedure that suited its interests. The invasion of the Falklands started on 2 April 1982. The Prime Minister announced that we were to send an expedition to retake them. We, in RAE, expected to be able to make a significant contribution in support of the military effort. Within 24 hours, I had received a memorandum from MoD to the effect that whatever we did was to be inside the budget already allocated. Whilst studying it and trying to think through our response, Don Hardy, Superintendent, Remote Sensing Division, Space Department, came into my office and said "We have just been asked to do the weather forecasting for the Falklands. If we are to do that task, I need a new computer. If it is ordered today, it will be delivered next week at a discount of 30%."

I felt that we had to do it and would sort out the implications later. The forecasting could be done via our own and American satellites and, even though we had no direct computer link with Northwood, the UK Operational Headquarters, we could somehow overcome that problem. In addition, by thermionic mapping, we could follow the movement of the South Atlantic thermal ridge and predict its position to the northeast of the Falklands. The ocean temperature could be measured in steps of 0.45 deg. C from 7.95 deg. C downward and give a clear indication of the variation of temperature with distance. Such thermal discontinuities were very important areas where the sonar capability is degraded and therefore could be a preferred area for the patrols of the UK Task Force during the campaign.

The FLIR and NVG programmes, described earlier, presented the U.K. with a fantastic night capability. None of these systems were, as yet, fitted to in-service aircraft and helicopters. To their great credit and benefit, the Services were very keen to use this brand new technology immediately. We set about changing the cockpit lighting on the helicopters and Harriers as rapidly as possible in the UK and then an RAE team sailed on the task force ships continuing the work until Ascension Island was reached. Several clandestine operations, to destroy Argentinean aircraft on West Falkland, were accomplished with the help of this equipment. In addition, there was a serious shortage of NVGs. We were able to purchase them direct from the USA and passing them, with operating instructions, not only to the helicopters but also to the Vulcans and Nimrods for in-flight refuelling in total darkness and to the Harriers for their night operations. New navigation and radio fix systems were rushed from the laboratory to equip the Vulcans and Nimrods.

Considerable work was done on the emergency fitting of air-to-air missiles to Nimrod and advice was given on the optimum use of our weapons and the means by which the effectiveness of enemy offensive systems could be reduced. In fact, we also produced a "laser dazzler" for use by against aircraft attacking ships in the Falkland Sound. Unfortunately, whilst it was being transferred from ship to ship, it was dropped into the sea and lost before it could be used. The rate at which we introduced these new systems at the start of the Falklands campaign was incredible.

There was an urgent need for missiles, with warheads installed, to be tested before being allocated the ships. Aberporth, because such tests were a normal trials requirement, was able to undertake the task immediately, at a cost of a considerable amount of overtime. The arming of Harrier with Sidewinder missiles was cleared.

The RAE wind tunnels were used to explore ways of putting up tents in high winds.

Each year all MoD apprentices competed for numerous prizes and, in turn, one group is selected to produce a present for the VIP presenting the prizes. RAE apprentices were, as usual, well to the fore in receiving these awards. It was known that Admiral of the Fleet Lord Lewin, Chief of Defence Staff during the Falklands Campaign, would be that VIP presenting the prizes and the apprentice School at Bedford chose to make a Coffee Table (or Sofa Table) based on a 19th century design. The Falkland Islands were the centre-piece and South Georgia and the South Sandwich Islands were on each flap. After the ceremony, Sir Frank Cooper, PUS in MoD, suggested that a similar table should be presented to Mrs Thatcher to recognise the part that she played in the South Atlantic Campaign. Andy Constant, who with three others had designed and built the table, met Mrs Thatcher and Sir Frank Cooper and presented it inside No.10 Downing St. Only those two tables were

The Prime Minister, Mrs Thatcher, Sir Frank Cooper and The Award Winning RAE Apprentice with a Falklands Table.

ever made and the drawings were then destroyed.

THE FUN

We were delighted to welcome many important visitors. Some came to be briefed on specific capabilities, others came to get a broad view of the work that we did. We always tried to make their day as worthwhile and pleasant as possible. Of course we had a message for every one of them. The more friends we had in high places, the better we were able to press the many changes that we planned. The most memorable visits were:-

Chief of Defence Staff, Lord Lewin knew us well from earlier contacts connected with the Falklands and the Apprentices. On this visit amongst many of the high technology programmes that we showed him, we took him to see a "direct voice operated cockpit". The cockpit displays and the navigation equipments were arranged to respond to voice instructions without the need for manual switching. It would also recognise the voice of the pilot and not respond to instructions from anyone else. At the time of the visit, MoD was in a very difficult financial position. Because of overspend, a moratorium had been declared on defence spending i.e. that we were not allowed to let any new contracts or pay any bills to contractors for the last six weeks of the financial year. In effect, we were taking a short term, interest free loan from our contractors. After he had given a series of instructions to the cockpit, I suggested to him that he might like to try "moratorium". To his considerable amusement, the cockpit responded with a "raspberry".

Viscount Trenchard, then Minister of State for Defence Procurement was briefed on the changes relating to the Strathcona Policy. He took me completely by surprise when, over coffee, saying that before he did anything, he would like advice on whether Princess Margaret should fly in a helicopter the following day. Once the day's programme was started, it took me a while to provide an answer by contacting the people who were really responsible. He visited Structures Department to see the Concorde Fatigue Rig and the use of Carbon Fibre Composites in aircraft structures as well as the work we were doing on aircraft battle vulnerability tests. He also visited Radio and Navigation Department to see the Comet Multi-role Flying Laboratory, which was carrying out a range of flying tasks in the field of navigation, communications and infrared measurements. Viscount Trenchard's family had a long association with the Establishment. In WW1, Lord Trenchard had based his headquarters in a building, near the Farnborough Rd, (used for the RAE Museum), before he flew off to France with a squadron of SE.5s. His father also visited, at the end of WW11, to see captured German aircraft and equipment. Viscount Trenchard was presented with a collage of photographs showing the Trenchard Building, an SE 5 and a group, including his father, visiting in 1945.

The RAE Apprentices

The RAE had an excellent apprentice school. Including those at the outstations, there were a total of nearly 500 apprentices under training (including 22 Chatham Dockyard 3rd and 4th year apprentices) About 330 were trained at Farnborough. The apprentices were not only very skilled craftsmen but they went on to take City and Guilds courses and to university. Every year there was a prize giving ceremony at Farnborough when a VIP presented the prizes, some of who were ex-apprentices. In 1983, John Farley, an ex-RAE Apprentice and Chief Test Pilot at BAe, presented the prizes. He said, " I am delighted to see that girls are well represented in the apprentice intake and had won some prizes. The bad news is that to be a success, the girls have to be better than the chaps. The good news is - that is not too difficult."

About every 5 years the Apprentices organised a reunion dinner. In 1983, the guest of honour was Michael Bentine, ex-goon, comedian and TV personality. He and his wife Clementina were guests at Farnborough for lunch and were given a short tour of the

By Kind Permission of the Aldershot News.

Establishment in the afternoon. He was no stranger to Farnborough. During WW1 his father, Mr. Adam Bentine, a Peruvian national, was a designer at Supermarine and a frequent visitor to the Royal Aircraft Factory. During the tour, I presented Michael with a model of Cody's Tree and, in one of the apprentice areas, Mrs Bentine received a posy of flowers. It really was a fun day.

He was a great believer in ESP (Extra Sensory Perception) and quoted examples of when it had worked. He also had a great fear that the Russians were working on mind control by the use of long wavelength radio waves.

After the Reunion Dinner he made a very amusing speech. He was assisted by a large scruffy bird called "The Ayatollah". He said, "Gentlemen, you are looking at the legionary "shee-ite hawk!" He also told some wonderful stories of the BBC attempts to sensor the stories in the Goon Show. They were always suspicious of the goon humour and never did understand stories about the "golden rivet". Anyone ex-service knew instantly what it was. After about an hour of hilarious fun, Peter Jowitt presented him with a "tensometer" of the pattern invented by his father in WW1.

ISM Awards

Every six months presentations of Imperial Service Medals were made. They were big occasions. As always, the wives and families were invited as well as the local press. It gave me a platform to brief the Establishment and the local public on the progress of the changes in RAE. The presentation of the awards was always a very happy occasion. In 1981, there were 20 recipients with a collective total of nearly 700 years of service to the Crown. The citations were read giving the history of their contributions and the medal awarded. The wives were each presented with an RAE brooch, made by the apprentices. After the ceremony, I had tea with them and chatted with most of them. It was always interesting to get their personal views on the Establishment and the changes being made. Some told stories of very amusing incidents in their lives. One of them had worked on some of the aircraft with me in Aero Flight some 35 years earlier.

The 75th Anniversary of Cody's First Flight at Farnborough. 16th October.1908.

We could not let such an anniversary pass without a major celebration. We were able to accommodate about four hundred people in the Assembly Hall. We invited about 70 VIP guests drawn from the local authorities, Senior Service personnel and senior MoD staff and members of Cody' Family. They started the evening in the Directors Mess for a Cocktail Party and then moved on to the Assembly Hall where 330 staff from Farnborough and Pyestock assembled for the evening performance.

I welcomed everyone and gave an introduction to Cody's early years. He packed at least four normal careers into his life, he trained horses, was wounded by Indians, prospected for gold, produced and rode in wild west shows where he made sufficient money to finance his experiments in kiting and flying.

Because of Cody's important show-business history, the second part of the entertainment was a review by the RAE and Farnborough Operatic Society on "The History of Farnborough."

I followed on with the first 25 years of Farnborough, including his achievements from the kite flights in 1901 to his tragic death on 7th Aug1913. This was followed by four lectures covering the next 50 years by:-
- DD (A), Eric Rogers on Airframes
- Head of Propulsion Dept, Frank Armstrong on Engines
- DD (E) Peter Whicher, on Flight Systems and
- DD (W) Derek Dawton on Weapons.

A 15-minute RAE film illustrating some of the recent work at RAE followed these lectures. Finally, I closed with a short review of "Where Next". At the end of the evening the Controller of Aircraft, Air Marshal Sir John Rogers gave the vote of thanks. We had also prepared an exhibition of Cody photographs. Everyone seemed to have thoroughly enjoyed the evening. We had some wonderful letters of thanks. It was a particular pleasure for the staff of RAE who seldom had the opportunity to get an appreciation of all that was going on around them.

The Samuel Cody School

The Wavell School Special Unit catered for about 80 children, aged 11 to 16 years, with moderate learning difficulties. It aimed to cover the normal school syllabus with the addition of more practical subjects. It had done so well that its achievement was to be marked by naming it the Samuel Cody School. I was invited to name the school and to mark that occasion, the apprentices had etched a special memorial plaque naming the school and showing the scene of Cody on his first flight over Laffans Plain on October 16,1908. This plaque was presented to Mr. Stuart Taylor, the Chairman of the Governors. The Head Master sent a very kind letter of thanks enclosing some 20 letters from the children.

The De Havilland DH100 Centenary Rally.
Tiger Moth Day at Farnborough. July 4th 1982.

It was 100 years since the birth of Geoffrey de Havilland. He sold his first aeroplane to the government in 1910 for £400 and went to work at the Aircraft Factory. After WW1, he founded his own very successful company. To commemorate this occasion and all the wonderful things that have happened in aviation in that time, some eighty-two Tiger Moths flew into Farnborough. It was romantic to see the tears in the eyes of some pilots- some even kissed the turf at Farnborough, some getting a logbook entry they never thought possible. What a pleasure it was to be able to greet them. Lunch was taken in the sunshine and no one stopped talking. Some parked their aircraft in front of the Control Tower, and others near the Black Sheds.

The COMET DH88. Refurbishment.

In 1934, the State of Victoria in Australia was holding its 100th anniversary celebrations and as part of that celebration announced a 12,300 miles Handicapped Air Race from Mildenhall in England to Melbourne in Australia for a prize of £15,000. De Havilland Directors were determined that the winner would be British and, even at a financial loss to the company, offered to build 200 mph racers at a nominal price of £5000 each. The offer was accepted and, with only nine months to go before the start of the race, three aircraft were ordered straight from the drawing board. Jim and Amy Mollison, Bernard Rubin and A.O.Edwards of Grosvenor House bought them. The latter aircraft, G-ACSP in a scarlet and white, was to be flown by C.W.A.Scot and Tom Campbell-Black.

The aircraft was a thorough bred racer, that placed heavy demands upon pilot skill and servicing know-how. It had two-position airscrews fitted that changed automatically to coarse pitch at 150 mph. A return to fine pitch could only be made on the ground, using a bicycle pump. That meant that the pilot could not go round again should there be a problem on the approach to land. The wing section was very thin, pointing to tricky stall problems at low speed; there was a split trailing edge flap providing some lift but mainly extra drag to give the pilot more control on the approach to land at a small airfield; the undercarriage was retracted and lowered manually; there was a tail skid that was low drag in flight but provided useful drag and directional stability when landing on grass fields; and it had brakes that could be operated separately by each foot pedal and could be locked-on together for parking. The skill of the pilots who raced in these machines must have been just incredible.

Despite many problems en-route, the Grosvenor House Comet won and landed in Melbourne in two days and twenty-three hours after it left Mildenhall. After the race, the aircraft was shipped back to the UK and used by the Air Ministry at Martlesham for trials with new enlarged air intakes fitted. It appeared in air displays and after an undercarriage failure during tests at high all-up-weight was put up for disposal as scrap. F.E.Tasker, who fitted Gipsy Six Series engines driving D.H.Hamilton variable pitch screws, saved it. In new blue and white livery, and renamed "The Orphan", it flew in the 1937 Marseilles-Damascus-Paris Air Race and the King's Cup at Hatfield. In 1938, renamed "The Burberry", it lowered the London-Capetown-London round trip to 15 days 17 hours.

In 1938, it left for Australia again under a new name "Australian Anniversary" but the undercarriage collapsed on landing at Cyprus. After repair it set off for Australia and New Zealand and then back to the UK, a round trip of 26,450 miles in 10 days 21 hours and 22 minutes to create a world record. Afterwards, with one engine donated to another aircraft, it remained under tarpaulins for 13 years until restored by students at the de Havilland Technical School in its original colours as displayed at the 1951 Festival of Britain Exhibition. In October 1965, it was handed to the Shuttleworth Trust for reconditioning. Many individuals and companies assisted the restoration work undertaken by the Shuttleworth Trust, but progress was very slow.

At the end of 1982, RAE was approached to see whether there was anything that could be done to help rebuilt it for display at the 1984 SBAC Show and the 50th Anniversary Celebrations in Australia. After careful consideration, we suggested that restoration work could continue as an apprentice exercise provided that there was no pressure for a completion date, that other companies would make available some of the necessary materials and that BAe Hatfield would help with the design work. Initially, all went well and there was some confidence in making these dates.

Once at RAE, the aircraft was thoroughly examined. Unfortunately, it was found that there was rot in the centre section as well as a fracture in the spliced joints. The original centre wing spars had to be removed, the rot cut out and new spruce spliced in place. There were numerous other problems with the planking and skins that had not been glued together correctly. To be sure that the aircraft would be safe to fly, we almost had to strip it down to its components and start the build again. It soon became clear that we were not going to be able to get it into a flying condition by the SBAC Show. We then made the decision that all the work done would be up to flying standard but the objective should be to have it in a beautiful ground display condition by August 1984.

We received a hint from Sir Peter Allen, President of the Transport Trust, that Prince Michael of Kent, Patron of the Trust, would welcome an invitation to see the restoration work on the aircraft. An informal visit was arranged for June 14th 1984. Within RAE, the credit for the work and organisation must go to David Cooke, the Chief Engineer, the Apprentice School, the Apprentices (both young and old), who devoted their time and effort to this refurbishment.

Our visitors that day included: -
Prince Michael of Kent
Sir Peter Allen	Transport Trust
Sir John Grandy, Marshal of the Royal Air Force	Shuttleworth Trust
Sir John Charnley	Shuttleworth Trust
Mr. Alan Curtis	DB Instruments
Mr. R.F.Kirkby	BAe
Mr. R.Roberts	BAe
Mr. Ron Paine Project Co-ordinator	Shuttleworth Trust

It was a particular pleasure to see Sir John Grandy again after 20 years and to welcome Alan Curtis, who was a friend of the Prince and had flown him into Farnborough.

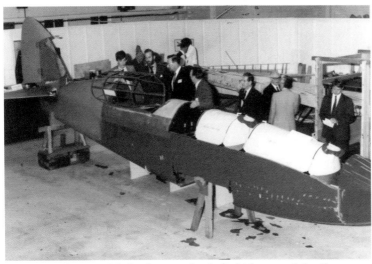

The Visit of Prince Michael to see the Refurbishment of the Comet.

After a thorough examination of the re-build, meeting many of those actually doing the work on the aircraft, Prince Michael dined with us. I was delighted to present him with a piece of the original aircraft, removed whilst rebuilding it, as a small memento of the occasion. It was a fantastic day for everyone particularly those who had done so much work on the project.

The Comet was splendidly displayed at the SBAC Show and was then flown, in pieces by Quantas, to Australia for their celebrations. It was then flown back by Quantas and went to BAe at Hatfield to be made flight-worthy for the Shuttleworth Trust. It is wonderful to see these historic aircraft in the air. Nevertheless, flying has changed. Today, it has become easier and

The De Haviland Comet.

safer. The tricks and skills that pilots used on the old aircraft had to be second nature. Because of improved design, they are no longer needed. But, put a modern pilot into a historic aircraft and he desperately needs to learn them again. My belief is that these aircraft should

- look as they did
- sound as they did   and
- smell as they did.

After that we should put in every safety device that would be effective in avoiding the risks that we might loose them to posterity.

<u>The Passing-Out Parade at the Gibraltar Barracks, Frimley.</u>

In March 1981, in a break from the tradition of senior Army officers carrying out these regular inspections at No.1 Training Regiment, Royal Engineers, I was invited to take the Passing Out Parade. I saw this as both an honour and a great pleasure and looked forward to the occasion. Despite my health problems at the time, it was such an unusual honour that I was determined to do it. The weather was freezing and there was a brisk wind across the parade ground. The very proud families of the cadets were there for the occasion. To start the programme, they demonstrated, on the parade ground, the military tactics that they had been taught. These demonstrations left no doubt as to their enthusiasm and the quality of their training.

By the time the parade finished, I was frozen and I have no doubt that everyone watching was too. I presented the prizes to the best all round recruit and to those winning the various prize competitions such as - best-improved recruit, skill-at-arms, physical training, and drill. Then to the tune of " The Magnificent Men in Their Flying Machines" I inspected the recruits. (Picture by kind permission of the Aldershot News). Feeling the intense cold both for myself and for everyone present, I suggested that I would speak to alternate soldiers and asked the C.O. if he would do the same. In that way

By kind permission of the Aldershot News

we would be able to talk to everyone. I thought that it all went well and we were soon able to retreat to the reception where the slow process of thawing out began. I tried to make up for failing to talk to everyone on the parade ground by chatting to all the parents with their sons over a cup of tea.

The National Gliding Championships

In 1984, Marconi Avionics plc, sponsored the National Open Class Championships at Lasham on 11-19th. Aug 1984. Peter Hearne of Marconi Avionics was their senior representative. He kept his glider there and did much of his flying from that airfield. The modern high performance gliders, made of fibreglass with carbon fibre composite structures, could achieve lift to drag ratios of the order of 50 to one. Such a glider can travel 50 miles from a height of 5000ft, without the assistance of thermals. Broadly it meant that whatever was visible could be reached and landed upon. Nevertheless, they are a real test of judgement on how long to spend in thermals, what water ballast to carry and, in particular, when to dump it. When they arrive back over the field, it makes a splendid spectacle to see them flying at high speed, dumping their water ballast over the finishing line.

These championships not only produced a National Champion but also made a considerable contribution to team selection for Britain's entry in the international championships the following year. For me the day was a great pleasure. Both Evelyn and I were

The National Gliding Championships. Lasham.

taken for a flight. I was thrilled by the occasion and certainly wanted to return for more flying but somehow there was never time.

Appointed Companion of the Bath.

In May 1983, I received a letter from Robin Butler telling me of the proposed award and that it would be announced in the June Birthday Honours List. I received over 150 letters of congratulation from the widest spectrum of people. In the RAE News, I said that I was delighted and honoured to receive the award. It was particularly gratifying that my family, friends and colleagues had been so generous in their congratulations to me. I am sure that the award reflected not only my own contributions to the aviation world but also by the RAE Board of Management and staff of the whole Establishment for their contribution to the considerable advances in technology and capability in the military and civil spheres. My family all attended the award ceremony at Buckingham Palace.

At the Palace with my family.

The RAE Societies.

There were about 40 Societies at RAE, most of which used the Staff Mess, the Assembly Hall or the Sports facilities for their events. I was President of the Bridge Club, the RAE and Farnborough Operatic Society, the RAE Symphony Orchestra and, for some 20 years, the RAE Ballroom Dancing Club. The Drama Club made excellent use of the Assembly Hall for their plays and always put on a very entertaining Christmas Pantomime.

Christmas Children's Parties

Every year, just before Christmas we had a party for the children between the age of about 7 and 13 years. Food and entertainment were laid on and at the end of the party each child was given a present from Father Christmas. They were happy affairs.

Christmas Tour.

Every year on the last two working days before Christmas, I tried to tour the Establishment, particularly to the parts that I had not visited during the year. At each stopping point, I was able to have a general discussion with the chaps about the plans for the future and their particular concerns. They always provided a drink. It was a tricky task to minimise the amount that I had to consume. The tour was always a race against time, but I always felt that every bit of it was worthwhile for them to see me and for me to get a flavour of the feelings at every level in the Establishment.

The RAE Messes

We were very lucky to have three messes within RAE. The Number 1 RAF Mess, which was run by the RAF but included the Navy and Army contingents and provided all the facilities of a normal service mess. Many civilians were invited to become Honorary Members and enjoyed that privilege. There were the usual social functions. In addition, on

SBAC show days, they had a small, tented encampment with grand stand seats from which to view all the action. Several times a year there were Ladies Dinning In Nights which Evelyn and I enjoyed greatly.

Quite separately, the Royal Navy celebrated their great battle victories. Each year they held a Dining-In Night to celebrate the sinking of an important part of the Italian Fleet by Swordfish (String Bags) at Taranto. A complete simulation of the battle was usually arranged above the diners with the battle ships represented by large polystyrene models in the middle of the dining hall. To the accompaniment of the engine noise and the sounds of battle, the Swordfish models, held on guide wires, flew over the top and dropped their "bombs" on the ships. Hopefully, at the right moment, the ships exploded showering all with polystyrene. I always attended these dinners in my father's dinner suit. It was turning a little green at the edges but it always went home white.

Scenes from my Farewell Dinner

The RAE Staff Mess was run for the benefit of all in RAE. It provided all the social club facilities for dances, snooker, annual dinners etc. A Mess Committee, chaired by one of the Heads of Departments, ran it.

The Director's Mess was one room with access to the bar as part of the Staff Mess complex. It was used to entertain special guests for lunch or dinner, for Heads of Departments to lunch in if they wished and for annual cocktail parties and farewell parties. On occasions when Heads of Departments were leaving or retiring, we held a party to wish them well. We always presented them with a small replica of Cody's Tree with two small slivers of the original tree inset into the base. I treasure the one I was presented with when I retired. Occasionally we presented one to an outside VIP who had helped us in our activities.

To mark my retirement, the Board laid on a super dinner, David Cooke taking the opportunity to expose some of the amusing incidents over the past few years.

It was a wonderful period being Director of RAE. Perhaps it was all summed up by a comment that a lady delegate made to me after I made the opening address at a Conference on the Use of Computers in Civil Aviation. "I cannot believe that anyone who looks like you and sounds like you would be Director of RAE. You should have cobwebs in your hair!!"

Chapter 11

Royal Ordnance and Hunting Engineering. 1984-1994

The "Diamond" Committee

In October 83, I received a letter telling me that my retirement would be on the day before my 60th birthday in June 84. After 4 years of service in the RAFVR (50% counted) and 36 years in the Civil Service, I was likely to retire on about £16k/year with a lump sum of £45k. It did not seem much and I felt far from spent. Having been in the defence business for 42 years, finding other employment that I might be good at seemed to be problematic. The letter also told me that *my attention should be drawn to the rules relating to the acceptance of certain types of employment within two years of retirement. As a guide, if I wished to take up any appointment with a business concern or other body which, in the broad, had any contractual relationship with Government, I was required to obtain the assent of the Ministry of Defence before accepting the offer of employment. ——After a lapse of two years from the date of retirement, the prior assent of the Ministry of Defence is not required. It should be noted that an application is also necessary for permission to take up appointments in the Service of a Foreign Government or international organisation.*

At the end of March 84, I applied to become a part-time, three days per week, consultant to British Aerospace. I also reminded them that I was assisting the Science and Engineering Research Council with their Manpower Committee and that the SERC might wish me to continue for some time. B.Ae. stressed that the job would involve the co-ordination of total company operational assessment tasks in the Aircraft and Dynamics Group and address the problems of the preparation of the company to meet the needs of the UK and NATO Forces in 2000 and beyond.

At about the same time, Fred Clarke, Chairman of the Royal Ordnance , approached me to become Director, Research and Development and to help with the preparations with the floatation of the company. However, there was a complication. The pension abatement rules governing retired civil servants who were re-employed with the government ensure that the re-employed civil servant shall receive no more by way of combined pay and pension than the salary that he was receiving immediately before his first retirement. If this limit was exceeded then the pension was abated by that amount, i.e. 100% tax. The effect was to limit my working time to less than three days per week.

The processes of the Business Appointments Committee (the Diamond Committee) seemed likely to be very slow. I had to make a decision in order to optimise the conditions of employment in the ROs and had to start the day after I retired from Farnborough.

At the end of 1984, Geoff Dollimore of Hunting Engineering invited me to join them as a non-executive Director and Consultant. Not knowing where the future was likely to take me, I applied through the MoD to the Diamond Committee for permission to take up that appointment.

Early in 1985, I received confirmation that I was not to be allowed to take up either appointment for two years. I felt that this was a harsh decision and sought permission to appeal against it. Lord Diamond agreed to let me appear before the Committee, which I did in April 1985. That was an interesting experience. It was suggested that I knew enough about the competitive position of most companies to be able to pass valuable information to others. I explained that I had been in that position for most of my career and, as I had not done so, it would be a major change for me to start. The result of the appeal was a continuation of the ban for two years for BAe and a reduction to one year for Hunting Engineering. The B.Ae organisational situation was in the process of change and they did not wish to hold the position open for another 18 months. Perhaps it was all working out for the best. I was then President Elect of the Royal Aeronautical Society, and would be President in May 1985. I was clearly going to have a very busy time.

The outcome of the review of the re-employment rules was published in March 1985. The main recommendations of the Commons All-Party Treasury and Civil Service Committee were designed to tighten the rules, by extending the maximum delay from two to five years on senior Whitehall officials and armed forces officers wishing to take jobs in the private sector. The Government rejected the recommendations. (What a stark contrast these rules were compared with those in other countries.) Although the committee had not found any hard evidence of impropriety, the ease of movement and the often informal approaches made by private companies meant that *"the traditional independence and impartiality of the Civil Service is in danger of being eroded in the eyes of the public."*

It is interesting to note that the rules applied only to those at and above Under Secretary grade. The people who really knew the new technology were below that grade. The arguments rumbled on. In 1989, the Commons Committee was pressing for further details of the operation of the Business Appointments Committee and the applicants. In supplying some 245 names without additional details, of those who had left the MoD between 1984 and 1988, the Government reminded the Committee that it *"had a duty to ensure the efficacy of the rules, which do not have the force of law, is in no way undermined"*. I felt that the whole thing was unjust. The problem was that those who worked by the rules were penalised and those who ignored them suffered no penalty.

Royal Ordnance

In 1560, the Royal Powder Mill was founded at Waltham Abbey. In 1984, ROs had four Divisions, Ammunition, Weapons, Small Arms and Explosives. The group sales were just short of £500m and the trading profit about £66m. During that year about 1% of turnover was spent on self-financed research and development, largely in aid of improved products in the areas of tank guns and ammunition, mortar bombs and naval mines. Historically, a great deal of the original concept work had been done in the Chief of Establishments and Research & Nuclear (CERN) organisation in RARDE Fort Halstead and at RPE Westcott before it was transferred to RO. As it was the only company capable of manufacturing and filling of explosives and propellants, the R.O. had involvement in the manufacture of most U.K. weapons and worked closely with a large number of companies.

Initially the R.O. operated entirely within government. It was then changed into an Agency, wholly owned by the government. At the beginning of 1985, it was established as a public limited company still owned by the government, with the view of establishing its market credibility before being floated on the stock market.

To have any hope of floatation, there was considerable work to be done. My terms of reference included: -
- responsibility to the Board for R & D policy and strategy
- the execution of that policy
- the delegation to the Divisions the appropriate projects, funding and time table and audit of these activities
- provide professional advice to the M.D. and the Divisional M.Ds on R & D
- generate the programme documents covering the work
- provide the HQ link with CERN
- be responsible for Company policy on Health and Safety at work.

I negotiated a Memorandum of Understanding with Chief of Establishments, Research and Nuclear (CERN). It was for continued research support, particularly from Fort Halstead, intended to cover a period of up to three years by which time it was expected that privatisation would be completed. R.O. would then be self-sufficient.

The next task was to establish a central Operational Analysis and Systems Analysis capability of the whole company. The nucleus of a group existed in some rented accommodation at the Royal Military College of Science RMCS at Shrivenham. The group had to grow in strength, its programmes and relationships with the Divisions established and the reporting of its work integrated with the reviews of work on current and new programmes in the Divisions. The location at RMCS was ideal. I obtained a long-term lease on some land within the Shrivenham complex. I was able to recruit some additional able staff for the unit and an integrated policy was established on the assessment of the performance of the company and competing equipment.

Rumour had it that the Chairman's view of an ideal HQ was that it was manned by himself and his secretary. It seemed that the best plan was to set up a research group in each of the Divisions and evolve the appropriate review procedures to ensure good programme control and communications within the R & D community and the company. I was equally keen to delegate down the Health and Safety task to the Divisional Directors of R & D, as they were likely to be the most expert in establishing good Divisional procedures.

By the spring of 1985, I had managed to recruit some experienced R & D Directors for the Divisions, John Brewer for Weapons and Fighting Vehicles, Dr Peter Lee for Ammunition Division. The other divisions had well established Research and Development capability- Explosives Division, particularly at Westcott, under Trevor Truman and Waltham Abbey under H Williams, and Dr David Izod at Small Arms Division. In Ammunition Division and Weapons and Fighting Vehicles Division considerable work was required to build up teams to a minimum level to render them viable. I then tasked these Directors with recruiting further staff. Luckily that was not too difficult because of the staff reductions at Fort Hallstead. By October 1985, we were able to produce a Corporate plan for submission to the R.O. Main Board. The aim of the paper was to take a first look at the Research and Development Strategy for Royal Ordnance and the most likely product ranges over the next 20 years. Its purpose was to generate a debate within the Company taking into account the technology, marketing and profitability etc.

In May 1985, I became President of the Royal Aeronautical Society. One of the very pleasant but strenuous tasks of the President is to visit the overseas branches. In July 1985, as part of that task, I visited Australia. I took advantage of being there to visit the Australian Ammunition Filling Station at St. Mary's, some 20 miles west of Sidney. It was a fascinating place. Vast by any standards, more than ten miles square with some 850 buildings and about 800 staff. In Canberra, I had long discussions with Professor Fink, Chief Defence Scientist and Fred Bennett, Head of the Office of Defence Production. They were very keen to hear of the policy in the UK because they had similar problems of overcapacity in Australia. At that time, they had simply by-passed the problem by creating artificial funding arrangements to keep the facilities ticking over. For any task, the Australian Services paid the market price for the production and another fund was found to make up the deficit.

To establish its long-term credibility, the R.O. needed to win some large overseas orders for defence equipment. That task had been made more difficult by the Ministry of Defence's new system of competitive tendering. By mid-1985, it seemed that that the flotation planned for June 1986 would be delayed. I did not feel that the floatation plan was feasible for at least three years and would be delayed from June 86 to at least 1989.

In November 1985, Coopers & Lybrand issued a report saying that, amongst many other issues, the policy on Strategic Defence Capacity would have to be solved before R.O. could be privatised. If Government wanted RO to become a fully commercial company before privatisation, it had to cut capacity to match foreseen demand unless there was an agreement with the Ministry of Defence on the need for a strategic surplus capacity to provide extra munitions and fighting vehicles in time of war. At that time the UK Government was R.O.'s biggest customer taking 85 % of its output. The Government wanted the privatisation to take place between that of British Airways and British Gas. Brian Basset, the Chairman, was determined to tackle the company problems especially in the management organisation, head office costs and search for new export markets for its munitions and armoured vehicles. In 1986, head office staffs were to be cut from 210 to about 30.

The earning restrictions limiting me to part-time could go on until privatisation was achieved. With the agreement of the Chairman, I looked for a replacement for me. As it happened, Derek Dawton, who had been my deputy at Farnborough, was retiring in February 1986. He was ideal and had the relevant experience. It would allow me to take up the attractive Directorship with Hunting Engineering. I joined them in November 1985 as a Director and remained a Director of R.O. until February1986 until I was replaced.

The Government decided not to wait until the RO could establish its market credibility and asked for offers to buy it. At least two consortiums made offers but a requirement to give assurances not to break up R.O. put off the consortium that included Hunting Engineering and it withdrew. British Aerospace acquired it but that is another story.

Hunting Engineering

In 1874, Charles Hunting purchased two wooden sailing ships and founded Hunting plc. Its interests spread into oil and aircraft and into the defence business. With the rationalisation of the aircraft and weapon industry in the late 1950's, the remnant of the Hunting Aircraft business was formed into Hunting Engineering. Geoffrey Dollimore was promoted through the company to become its Chairman. He guided it through its growth assisting RAE and other establishments with the defence work on both conventional and nuclear weapons. In the 1960s, RAE was attempting to put more of the design and development work out to industry. I had known the company since the mid-1960s and had a high regard for its all round capabilities, particularly in concept, analytical assessment and project management. HEL manufactured only about 20% of its products, the bulk being sub-contracted to other companies.

It was a particular pleasure to be joining them as a Consultant and Non-Executive Director. Tom Grievson, the M.D., had a clear view of the contribution that he wanted from me. He wanted me fully involved with the work of the company and, at the Management Boards to be sure to ask those difficult questions that helped to highlight our strengths and take action to overcome our weaknesses. In February 1986, the Chairman invited me to become Technical Director. That appointment was put off until June when I became free from the restraints placed on me by the Diamond Committee.

Whilst I was Technical Director, the Systems Division and Engineering Divisions reported to me. In the former, the work covered Conventional weapons, Special weapons, Warheads and Terminal Effects and Mission Analysis Departments and in the latter, Projects, Weapon Products, Micro Systems products, Vehicle Products and Engineering Support. We were involved in a considerable variety of studies and in bidding for and producing a wide variety of weapon equipment for the RAF and Army and, to a lesser extent, for the Navy. I will describe some of the programmes in support of the three Services and some of the complexity of the political and funding problems of these projects.

Land Weapons.

The development of LAW 80 (Light Anti -tank Weapon), below, had started in the late 70s. RARDE and RPE had done the concept work and early research packages. The concept brought together three important technologies: -

LAW80

a) a fast burning rocket motor that developed its full thrust and completed its burning within the length of the launch tube (very important for the protection of the firer). There was no reaction on the weapon during firing.

b) a shaped charge warhead that could penetrate over two feet of rolled homogeneous armour (below) and

c) a spotting rifle that had five rounds of *tracer* ammunition to assist with the aiming particularly against crossing targets.

The weapon weighed about 9 kilograms and was still serviceable after it had been immersed in water for 24 hours. Arming of the missile was delayed until it had flown more than 20 meters from launch, very important for the safety of the firer. HEL and R.O. produced it. It was an excellent weapon well liked by the Army with considerable potential for overseas sales.

LAW Warhead Effectiveness.

It was in head-to-head competition with the French, APILAS. The difference between the two weapons was interesting. LAW was designed as a short-range weapon with a very simple, disposable sight combined with the tracer spotting rounds. The infantry purchased eighty thousand of the 140,000 ordered. In their mode of operation, tanks are engaged from the side and from the rear. In most scenarios, the British army would expect to have medium range weapons, such as Swingfire, taking on the targets at longer ranges.

APILAS was a longer-range weapon, with a sophisticated magnifying removable sight that gave higher accuracy at longer ranges. The efflux from the rocket motor, which was still burning when it left the tube, required the firer to need protection with a facemask. We had photographs of the operators with burning fragments on their clothing after firing. The French also choose to use a specialist unit to demonstrate their weapons, whereas we used the serving soldiers who were to use the weapons.

Each of the companies wished to demonstrate their wares to the greatest advantage. We would suggest demonstrations at shorter ranges with some crossing targets whereas the French opted for directly approaching targets at the maximum range of their weapon. It is then a matter of customer judgement to choose which weapon suited them best operationally and who offered the best deal. As with rifles, there is a grave tendency in peacetime to veer towards the accuracy at the longer ranges. It is at these ranges that units compete both nationally and internationally, rather than in more representative operational conditions.

By the 1980s, research into the technology of the countermeasures to shaped charge was developing. In the UK, we had moved onto Chobham Armour, a sandwich system of armour and composite materials. The Russians had introduced ERA (Explosive Reactive Armour) designed to disrupt the plasma jet from the shaped charge. When the jet reaches the explosive sandwiched between two plates of steel, it detonates, throwing the plates apart, thereby disrupting the plasma jet and reducing its length and therefore its effectiveness. The effectiveness of the jet in penetrating armour depends upon the maintenance of both its straightness and continuity. The explosive sandwich plates were fitted inside metal boxes, like lunch boxes, in either single or multiple layers. They could retrofit these active armour arrays to the front, sides and top of the tank and turret to counter attack by surface and air launched weapons. It is probable that the Israelis first fitted this armour (Blazer). When the Syrians captured some tanks in the fighting in 1982, it is likely that they passed the technology to the Soviet Union. These armours were very successful in severely degrading the effectiveness of the single shaped charge warhead. The reactive armour was cheap, ~ 6% of the cost of a tank, about 1/3 of which were retrofitted. Unfortunately, these boxes could be detonated by other means i.e. bullets, and therefore there was some loss of safety to the crews.

To counter this improvement in tank defences, John Nicholson and his team in the Warhead Group studied a variety of tandem warheads. The British Army wanted to see a tandem warhead fitted to LAW but they did not see the need for this modification until at least 2003 and had no provisions in their budget for this update. With some support from both MoD, the Defence Advanced Research Programme Agency, DARPA, in the USA, and a considerable input of company funds, the research work to up-rate the warhead went ahead.

The options were to: -

a) design a tandem warhead. The first warhead would detonate the ERA plates. They would then fly apart. By delaying the detonation of the second warhead until the plates had moved out of the way, it could then penetrate the armour of the vehicle, or

b) design a tandem warhead in which the first warhead produced a hole in the explosive plates *without detonating them*, so that the jet from the second warhead could pass through that hole in the ERA undamaged and attack the target vehicle behind.

The design problems of either solution were very difficult but, if the latter could be made to work, it appeared to have the greater merit.

To assess the overall lethality of these weapons, very complex target description models have to be built. We felt that to extend the LAW technology for one further generation was both economic and desirable for the British Army. Nevertheless because of the funding problem, it seemed likely that they would procrastinate for some years.

Adder

In the meantime there were derivatives of LAW 80 that could fill other military needs. A typical example was ADDER. It was a LAW 80 mounted on a tripod that could be activated remotely. There were three variants, a remote firing system by cable or radio making it safe to fire from an enclosed space, or by a sensor system to detect the target and initiate the weapon.

There were other opportunities to compete in both experimental, proof of principle and development phases of other programmes. ACEATM Aimed Controlled Effect Anti-Tank Mine (or ARGES) was a European international collaborative programme. We were working with Dynamit Nobel and Honeywell of Germany, and Giat of France. Our responsibilities in the project were the launch tube, the stand and the warhead. The concept was not dissimilar to ADDER (above) but was intended to be a next generation weapon effective against the sides of tanks fitted with the latest Explosive Reactive Armour defences. Something like 20 different targets were defined each having multiple explosive sandwiches within them. It required considerable ingenuity to devise and trial numerous warheads against these targets. It could not have been done without the Hunting sophisticated trials facilities at Porton Down. The problem with this programme was that despite the assurances of a large purchase, particularly from Germany, I never had any confidence that it would ever be funded beyond the demonstration phase. If I was right it would be a poor commercial proposition but it might be a price worth paying to keep the lead in our own technological strengths.

We also had a contract from DARPA in the USA for research into warhead design that was effective by using the non-detonation technique.

The Army had been seeking a barrage rocket system to update its artillery capability. It chose the Multiple Launch Rocket System MLRS and Hunting were part of the management consortium. It was used very effectively in Desert Storm.

Air Weapons

Hunting Engineering had a long history of successful development of air delivered weapons. In the early 1960's, they concentrated on the development of the WE177, the nuclear retarded bomb. In the 1990's, Hunting were also involved in the decommissioning of that weapon whilst they were managing AWE at Aldermaston.

In the late 1980's, the one of the top requirements was SR(A)1244, the Future Tactical Nuclear Weapon FTNW, a stand-off cruise weapon. We were working with BAe on an all-British solution. After some years of work, the requirement was dropped in favour of reliance on Chevaline.

Hunting had also been successful in bidding for the retarded bomb, the B10, and BL755 and JP233 (both covered earlier). In the 1990's, these systems were still very much in-service and needed tests, trials and refurbishment. Improvements to BL755 were proposed to the RAF to make it suitable for release from altitudes of 10,000ft or higher. There was considerable overseas interest in the weapon, but buyers are cautious if it is not in service with the national air force. At the time, the RAF reluctance to acquire it appeared to rest on their view that their survival depended on operating at low altitude. Because of the overseas market, we offered to give the RAF a stock of weapons so that we could rightly claim that it was in-service. Their reluctance continued. The French would appear to have had no such qualms.

SR(A)1236. This Staff Target was a requirement for a long-range standoff missile for use against hard targets such as command posts. When I joined Hunting, there was an international programme in progress called LRSOM, Long Range Stand-Off Weapon. After the bids had been put together, this programme seemed to disappear into the sand and be replaced by another international programme called MSOW, Modular Stand-Off Weapon. Initially three international consortia were formed to bid for the development and production of this weapon. Eventually one consortium dropped out leaving two. In our consortium, nine nations and ten contractors were involved. The competition rules required each element of the weapon system to be competed within the consortia. The share of the work was in proportion to the likely national buy of weapons. The resulting weapon systems were then competed between the two consortia.

As part of the long-range version of MSOW, Hunting contributed their expertise on

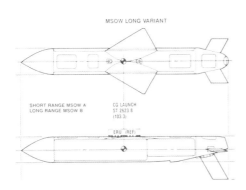

dispensing systems. This missile was designed to carry a wide variety of payloads. The programme was for a Project Definition phase from mid 88 to the end of 1990, full-scale development, 1991 to 1995 and production, with some overlapping development, from 1995 to 2000. I was always concerned that the programme would take much longer than planned as no time had been allowed between the programme phases for the governments to agree the next phase; it looked like a very long programme.

The Modular Stand-Off Weapon. MSOW.

There was always a concern that the USA, one of the main participants, had a "black" programme that would fulfil these roles. The project definition programme, funded by the companies, was tortuous. From time to time countries dropped out of the programme. On every occasion, the workload had to be redistributed between the remaining participants down to within one percent. In all there were eleven invitations to tender, each involving a considerable amount of work and cost. Eventually the final bids were submitted. After a lengthy period, the loser was declared. The US government then decided to withdraw and *no contract* would be awarded to the winner.

The RAF still had a requirement for a stand-off weapon for the attack of airfields, of other key points on land and ships and other targets at sea. Unfortunately they did not define the requirement as clearly as usual except that they wanted a unitary warhead. We expected 1242 to take care of the short-range requirements. We then searched for possible partners in Europe and the USA. There were plenty around. There were at least eight vehicles that might be developed, at low cost, to meet the requirement. We had looked at the French Apache several times but doubted whether the armour-piercing warhead was adequate for penetration of the tough targets at the speed of the missile impact. Even more worrying, was whether the price could be brought down to an affordable level. In the end, we favoured the US Navy Weapon SLAM, Stand-off Land Attack Missile, with extended range and modified to meet the RAF requirement. Without doubt, it was an excellent competitor.

B.Ae. chose to work with the French on Apache. They extended its range and then made two very important changes to create Storm Shadow: -

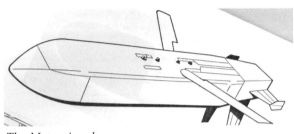

a) they stretched the meaning of "unitary" warhead to include three warheads. The combination was called "Broach".

b) somehow they induced MoD to get involved with the French Government in an attempt to reduce the price of the vehicle to a

The Matra Apache

level that was more in keeping with the UK affordability and the competitive environment.

I am sure that the French, using some marginal costing technique, achieved the cost reduction. It was a very smart move by B.Ae. The contract went to them. It is also possible that a moment of "go European" might also have slipped into the decision equation.

SR(A)1238. The RAF wanted a weapon that would replace BL755, the anti-tank weapon, for delivery from aircraft, particularly the Harrier, for use in close support for the Army and against follow-on support forces being brought up from the rear to the battle area.

This programme stretched from the late 1970s through to the mid-1990s before a contract was let. Over that period, the threat changed from "wall to wall" tanks in a major conflict between super-power opponents, NATO and the Warsaw Pact, through to the collapse of the Soviet Empire. Major technological advances and large reductions in the defence budgets to match that reduction and redefinition of the threat paralleled these events. However, the targets, though reduced in numbers and redistributed in location were still as difficult to destroy, because they were likely to be the same targets.

In the late 1970s, SR(A)1227, called for a replacement for BL755 the cluster weapon, designed for attacking tanks from the air. The solution proposed at that time was another unguided cluster weapon using a shaped charge to attack the top armour of the tank. Should it not impact a tank or similar vehicle, the weapon would sit on the ground and act as a mine against the same threat.

This concept was not viewed as adequate. Terminally Guided Munitions (TGMs) were becoming technologically possible and could be delivered with great accuracy and achieve a greater number of kills per mission. In overall terms, the cost/kill would be cheaper. RAE led these studies. A typical TGM would have folding wings and be terminally guided to the target

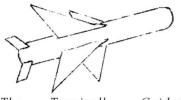

The Terminally Guided Muniton. TGM.

by either radar or IR seekers. In 1982, two consortia, one led by B.Ae., the other by Hunting/Marconi, were set up to compete for this weapon system. Feasibility studies were started in 1983.

The RAF had historically provided Close Air Support (CAS) for the Army at the forward edge of the battle area and, in depth, Battlefield Air Interdiction (BAI) by attacking the enemy ground formations well behind the front.

In December 1985, the Equipment Procurement Committee endorsed two funded project definition studies leading to a design using a dispenser carrying eight TGMs. At that time, I was responsible for running a "Red Team Review" of the consortium proposals before finalising the submissions to MoD. Having been sustained through the day by sandwiches and coffee, the whole exercise was concentrated within one day, starting at 9.00am and finishing about 9.00pm. I rang Evelyn to tell her that I would be home by 10.00pm. Unfortunately, an oil tanker was on fire in the M25 and I did not get home until well after midnight.

The next day I had a long discussion with Tom Grievson, the MD. I was able to tell him the good news that the whole review had gone well and, although there was some more work to do, we would be ready with the submissions on time. I also told him the bad news that the whole concept seemed to me to be too expensive and it was unlikely that anyone would buy it. After he got over the shock, he asked what we should do. I suggested that we should take a small team from the project group and task them with finding a cheaper solution using technology or equipment that might already be available in Europe or the USA. The whole exercise was to be kept strictly "company confidential". I did not endear myself to the project team with these proposals. They had been working on the programme for many years and were rightly proud of their solution.

The bids were submitted to MoD, the documents travelling to London, in the time honoured way, of a large number of copies in a lorry and two copies in a separate car, thereby ensuring that the bids arrived before the deadline.

The MoD analysis and comparison of the bids would normally take six months. In that period many supplementary questions would be asked and answered. After nine months there were no signs of a decision. If your bid is chosen, then there is great pressure to complete the work within, or even earlier than, the offered timescale. To do that the project team has to be built up and this "marching army" needs tasking and paying. Still no decision was emerging from London.

We met BAe quite regularly to discuss other programmes on which we were working together. Over lunch or dinner, 1238 was discussed. We were both in the same difficult position. Both companies decided to press MoD to make a choice or cancel the programme.

In MoD, Sir Peter Levene had replaced cost-plus contracting with fixed price contracts. These meant that contractors were responsible for cost overruns and would lose financially if the equipment did not work or was not delivered on time. The contractor responsibilities had changed but the MoD had not matched this change with the improved mechanisms required for managing in this competition environment. Perhaps it was not surprising that progress was slow or non-existent.

Continued pressure from both companies upon MoD to make a decision eventually led in February 1987 to two (under) funded six-month studies by the two consortia i.e.the original proposals, with the view to reducing the price. This programme became known as the "compliant solution". In addition, an open international invitation was made for submissions of un-funded lower cost solutions that did not meet the original stringent requirements, hereafter known as the "non-compliant solutions". It was MoD's commitment to the programme that enabled UK companies to seek the developed technology from overseas contractors that might generate an affordable solution.

In October 1987, MoD received the bids offering the two "compliant" solutions and six "non-compliant" solutions. The intention was to choose one solution for fixed price development and production.

Hunting/ Marconi submitted a "compliant" bid together and a separate "non-compliant" bids from each company. It caused some tricky liaison problems. The Marconi technological input to their "non-compliant" solution, i.e. the seeker system, was probably the same technology in their contribution to the TGM seekers in the joint "compliant" solution.

Consequently, they were very reluctant to provide the cost information for the Hunting/Marconi bid. The bids were submitted on time but only just. The overall price difference between the "compliant" and "non- compliant" bids was large.

Of the six non-compliant bids, the RAF soon decided that the only two that appeared to have potential were SWAARM and BRIMSTONE. The RAF faced a dilemma-the effectiveness that they required was unaffordable and the affordable solutions fell below the effectiveness they thought that they required.

The compliant solution had consisted of a dispenser carrying eight TGMs. In the short-range mode, the aircraft flying at a low height released the dispenser. It was then decelerated to fall behind the aircraft and start a climb. The TGMs were then ejected to start a rapid programmed climb to about 150m high in level flight. The seeker then started its search for targets so that the first could be attacked at a range where the target could be identified. As the TGM glided and slowed, the search would continue until a target was acquired or the TGM slowed until it stalled. The coverage from one dispenser was some 900m long and 400m wide.

SWAARM and BRIMSTONE
Publicity Mats

In the long-range mode, selectable in flight, the dispenser would pull up behind the aircraft and glide up to eight km., the target area being fixed by GPS etc and a radar/IR detector for starting the TGM ejection sequence.

When offering a "non-compliant" solution, early decisions have to be made on which of the original requirements needed to be fully met or could be relaxed. The design of the "compliant" solution was driven by the requirement for the minimum range. In the close support role and operating at low altitude, this seemed to be vital. It enabled the pilot operating alone to search, acquire and identify the target, aim and fire. The task would be easier with forward air controllers (FAC), perhaps using laser markers, but a short minimum range seemed vital. SWAARM was designed around this philosophy. The sixteen sensor-fused munitions (SFMs) were laterally ejected in opposing pairs to cover an optimised ground pattern sufficient to cover a tank column.

Following ejection, each sub-munition is retarded into a controlled vertical descent in two stages. Firstly, a single cruciform retarder is deployed to reduce the sub-munitions speed.

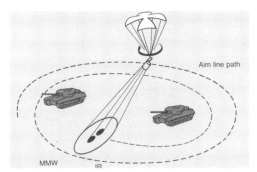

The Sensor Fused Munition Search Operation

A vortex ring parachute is then deployed which stabilizes the sub-munition body offset to the vertical to enable the dual-mode Milli-Metric Wave and Infra Red system to describe a decreasing spiral search pattern on the ground. This dual seeker system provides high resistance to countermeasures. When the sensor system detects the target, the warhead is detonated and a solid high velocity Explosively Formed Projectile (EFP), a slug, is fired at the top of the target. A double layer of explosive reactive armour might provide the protection for a typical target.. Travelling at many times the speed of sound, this slug can perforate the ERA protection and the armours of future Main Battle Tanks and generate behind armour effects after punching a large diameter hole in the target. It was being developed for a US artillery system, thereby saving a great deal of the development costs and, by increased production, lowering the price.

Damage to Top Armour.

The Marconi solution was to use an extremely successful US missile, HELLFIRE with the Marconi radar seeker installed at the front. They had arranged them to be carried in a group of four under each wing pylon. At the time, my assessment of it was that the weapon had no hope of getting anywhere near the minimum range requirement and the under-wing installation of four missiles would shake itself and the aircraft to pieces at 450 knots.

In August 1988, MoD announced that the Hunting/Honeywell and Marconi/Rockwell proposals had been selected for continued work on a risk reduction programme to lasting for two years. In autumn 88, each company made proposals for the risk reduction programmes. For Hunting, it was further work on the warhead and for Marconi it was further work on the seeker. Assurances were given that these programmes were of the highest RAF priority.

In January 89, negotiations to reduce budgets on the bids for the risk reduction programmes continued. It was expected that the programmes would start in mid 89. A full team would be required for the start of the project and a major recruiting programme was started to bring the team up to strength. The team has to do something and the work on the risk reduction programmes was started at company expense. In June 89, we were requested to comment on a press release to be issued by MoD on signing the start of the Risk Reduction Programmes. By the end of 89, we were getting two conflicting messages. On the one hand, we still felt that there would be a risk reduction contract. On the other hand, MoD PE were emphasising that there was no commitment prior to a contract award, as is required of them. Rumours from the centre of MoD indicated that some felt that, as they were getting the risk reduction work free, why did they need to give a contract.

Meanwhile the world around us was changing. In 1989, the Berlin Wall came down and the Soviet Empire was beginning to crumble. There were serious questions on the defence needs of the future and the Peace Dividend called for a reduction in the defence budget. We had great worries that the MoD study called "Options for Change" and reduced budgets would affect the placing of a contract after such a considerable period of study at company expense.

The risk reduction contracts were never funded and the companies funded all the work. In Hunting and Honeywell, the effectiveness of the warhead against a number of possible tank top-armour concepts was demonstrated.

After I had retired, it was announced that Brimstone was to be ordered. At the SBAC show of 1994, Marconi exhibited their solution. The interesting changes were to the under-wing installation. The number of missiles had been reduced from four to three in a much cleaner aerodynamic installation. This change would mean a reduction of 25% in the effectiveness of the pylon load. They claimed that this reduction had been recovered by increasing the sweep rate of the radar, covering the search area for targets more effectively. For a variety of reasons, I believe that MoD chose the wrong solution.

However, 1238 will be a wonderful exercise for the military staff colleges to look at the changing threat, the requirement drift, the role that UK might play with its allies in a future war, the politics, the financial constraints and the operational circumstances in which the weapon might be used. For industry, it was a painful experience. In a wider context, the President of the Grumman Corporation expressed it by saying, "if we allow competition to turn into a contest to see who can bleed the longest, no one wins and we only drain the strength of the industry."

In Hunting, we had many intense debates to try to identify the programmes that would be funded and that we could win. Brian Hibbert challenged me to predict which programme was the most likely to be part of the future MoD programme. My prediction was the Westland Apache Helicopter programme with the Longbow/Hellfire weapons fit. He was a very successful in making HEL part of that winning consortium.

Nuclear Weapons.

From the earliest days when the first team was set up under Dr. William Penny to develop the UK atom bomb, industrial help was required to assist the government teams to carry through their task. As one might expect, the responsibilities were: - AWE for the nuclear package, RARDE for the H.E. package that provided the energy to start the nuclear chain reaction and RAE to design the bomb casing and other related issues. RARDE in particular wanted equipment to handle the sensitive nuclear package and Hunting contributed to this task. Thus began a long association with the Establishments on the first UK weapon, Blue Danube. Hunting not only set up support units in Farnborough but also at Woomera range in Australia. This association continued through, Blue Steel, WE177, Polaris and Chevaline. The System Assessment part of Hunting was involved in the missile-missile battle and work related to Star Wars. Thus the interactive war-game became part of the Hunting expertise and workload.

Studies were made into the performance of the Iraqi extension to the Russian SCUD missiles intended to increase the range to enable them to attack Israel. One of the predictions made by the team, led by Neil Cox, was that the missile would break-up during the violent longitudinal oscillations as it re-entered the atmosphere. The TV pictures transmitted from Israel showed two things quite clearly. Firstly, that the body of the missile broke up by pulling the rivets across the centre section and secondly, that there was no obvious damage caused by the attack by the anti-missile system deployed to defend Israel and help to persuade it not to enter the Gulf War. The Patriot firings were spectacular and Israel did not enter the war.

The Atomic Weapons Establishment.

AWE was unique in the world. The whole UK military nuclear industry was located there. Many felt that the vital continued support from the USA depended upon AWE being totally operated and manned by government servants. The work was divided into: -
a) R & D, where it had to be demonstrated that we knew the science and were making progress in the technology of nuclear weapon before meaningful collaboration could be achieved. To do that, the work had to be world class.
b) the site operation of a major nuclear facility.
c) manufacturing and weapons and components and
d) the support and eventual decommissioning of the Service weapons.

From the mid 1970s, AWE was short of manpower and moves of staff from other Establishments, particularly RAE and NGTE were quite regular. It had not overcome the recruiting and retention problem, probably due to inadequate pay in a very affluent area. It was becoming clear that staff shortages would cause a delay in the weapon production programme. There was also a major problem of delays in the £1 billion capital programme that could be attributed to the Property Services Agency and the Contractors as well as AWE.

The Prime Minister, Mrs Thatcher, invited Sir Francis (now Lord) Tombs, Chairman of Rolls Royce, to examine AWE and let her have his recommendations on the changes required. He recommended that the R & D should remain under the Crown for the time being but the production, experimental manufacturing and engineering should all come under a contractor and should be in the private sector thereby removing the restraints on management freedom and employee terms and conditions. MoD felt that the timescale recommended was impractical and that the extreme solution would create severe industrial problems. This possibility, in a situation where safety considerations were paramount, was highly undesirable. In addition, any solution would require legislation and that would take time. It was also vital that there should be no conflict with the 1958 UK/US Mutual Defence Agreement.

In the USA, on programmes of high national importance, Government Owned, Company Operated (GOCO) facilities had been used successfully for many years. Some of these facilities were operated on a cost-plus-award-fee arrangement in which the contractor receives a special fee based on performance. The US model was attractive to the UK but the US had no experience of changing the status of an existing government laboratory to a GOCO. Therefore if that policy was chosen, there would be complex problems to solve and many issues to be settled. In such a change, it would be vital that everyone, the managers in MoD, the contractors and the staff at the laboratory, be committed to the programme whilst the whole process of change took place.

In December 1989, the Secretary of Defence made a statement in the House of Commons on the future organisation at AWE. In that announcement, he proposed the GOCO model be adopted and that the contractor should bring in a small number of experienced top managers and initially concentrate on manufacturing and site support. Under the initial arrangements, AWE employees would remain civil servants. It would then be followed by another contract for full contractor operation. The contractors would be selected by competition and invitations to tender would be issued as soon as possible. The invitations were issued in January 1990 in the MoD contractors Bulletin.

The Hunting involvement in the nuclear weapons programme over many years put it in a key position to make a successful bid. Some felt that RR had an inside track because of their experience of running the Navy nuclear shore facilities and the earlier study by Tombs. Tom Grievson put together a consortium led by Hunting Engineering, expert in project management and production organisation, a US firm-Brown and Root, very experienced in developing and operating large facilities and the Atomic Energy Authority expert in nuclear manufacturing and safety. The new company was called Hunting-BRAE.

There were some interesting little twists in the run up to the award of the contract. At one stage, it was thought that the contractor management team might have to become temporary civil servants to ensure that Ministers were confident that their instructions would be followed. A new relationship would have to be developed with the Nuclear Industry's Inspectorate and the conditions for the contractor might be difficult. To monitor the performance of the contractor, a Compliance Office needed to be set up within the

Establishment and its role and approach had to be established by experience. With our fingers crossed that everyone was committed to the success of the new arrangements, Hunting-BRAE bid and won the Phase 1 contract. It was very successful.

There was a delay in the competition for the Phase 2 contract. In that contract, each area of work would be contracted separately and there was much to clarify in this multiple contract relationship with MoD. In a situation where expertise was scarce, the peaks and troughs of the nuclear research and development tasks had historically been smoothed out by allowing some of the resources of the research activities to be moved to development and vice-versa as the commitments varied. It was important to establish how this might be done under the multiple contract system. There was also concern that significant changes in the organisation could be required that would be difficult to do without a very clear indication of the forward programme. The Phase 2 part of the competition was again won by the Hunting-BRAE. It really was a great achievement by the consortium.

Somehow, the successful and very stimulating work on ballistic missiles, anti-ballistic missiles, Star Wars and Nuclear Weapons more than compensated for the frustrations of the work on conventional weapons.

Diversification and Other Interests.

Alchemist. It was clear that the Defence market would be reducing over the coming years. The company looked for other markets in which to sell its skills. This task is incredibly difficult. One of the proposals was to enter the crop spraying business by producing a new high technology sprayer in conjunction with the Silsoe Agricultural College. It was called Alchemist. It was to be designed to accurately meter four different chemicals into water at a point between the tank and the spray boom. The intention was to make it a closed transfer system to take the chemicals direct from their original containers without operator contact thereby making it safer for the operator. There would be no chemical wastage or spray to dispose of at the end of the day and the operator could introduce the chemicals on the move for "patch spraying" problem weeds such as black grass and wild oats. It was to be marketed by Gem Sprayers Ltd at about £16,000/copy.

A combined project team made a presentation to the senior management. The business plan predicted that there would be no profit for more than six years. I was very unconvinced by the proposal. I visited Silsoe and Gem Sprayers (with my brother who had just bought a new sprayer from them). I also had a long discussion with the Agricultural Engineering Association. In an article in Arable Farming, it said, "despite the already overcrowded market, sprayer companies took the opportunity to introduce more new spraying equipment than ever before. This was despite the fall in the market from 6000 units to 2000 units/year over ten years". After further consideration by the Company, it was not surprising that this programme ceased to be funded.

Stays at Bedford. Bedford was a long way from home and to relieve the travel pressure, I stayed many times at a Hotel/Health Farm called Mallets. I ordered a prime scotch fillet cooked in wild mushroom and artichoke sauce so often that the Chef decided to put it on the menu. Despite it being the most expensive main course, Tournados Kerr became a best selling item.

Farewell to Hunting. In late 1993, a number of factors came together to convince me that it was time to retire. In 1994, I would be 70 and despite enjoying the challenge of the work, it seemed unwise to go on longer than that. Many of the programmes that we were competing were hung in the air because of changing threats, shortage of funds and a failure to recognise that if we were to compete then the programmes had to be more than national because our home market was too small. Tom Grievson agreed that I could retire at the end of March 94 if I found a replacement for myself.

There were many farewell parties. Tom Grievson presented me with the cartoon signed by my colleagues depicting the HEL building complex as "STALAGLUFT 1" with Tom Grievson calling from the window "Come back Tom", whilst I am digging my way under the

Retiring Cartoon.

security fence to escape with my bus pass and retirement books. The Directors were also very kind in the things that they wrote in their farewell messages. The highlight of this sequence of events was the farewell party at the famous hotel Le Manoir aux Quat' Saisons. near Oxford. Evelyn and I were given the best suite that included a double Jacuzzi. Before breakfast, we decided to bath together. I emptied the bath soap bottle into it. The foam climbed way over the sides and all surfaces were so slippery that we were, amidst considerable laughter, unable to get out. The bubbles were billowing from the drains past the dining room window much to the amusement of all dining and to our embarrassment when we eventually arrived down for breakfast. So ended a fascinating career in which Evelyn has always given me 110% support.

HUNTING ENGINEERING LIMITED

GENERAL NOTICE

MR TOM KERR CB

Mr Tom Kerr, who is much older than he looks, will retire from the Board at its next meeting and retire from Company service at the end of March 1994.

Tom Kerr has had a very distinguished career within the Civil Service and a long and happy association with the Company.

His CV starts with a 1st Class Honours Degree from Durham, progresses through to pilot and flying instructor, an engineer at RAE on aero flight, Head of Supersonic flight and then management of weapon aspects. By the late 1960's he was Deputy Director at DOAE and then, several jobs later, in the late 1970s the Director of National Gas Turbine Establishment. In 1980 he became the Director of RAE and was responsible for all of the work of the establishment including of course, the staff, some of whom had probably been beastly to him when he was a young engineer there many years before!

He retired from the MOD in 1984 and, following a brief spell in Royal Ordnance – about which he has little to say – joined us in 1985 as a Director and consultant. In June 1986 he agreed to be our Technical Director following the departure of the previous incumbent and pending the arrival of a replacement. He did an excellent job to the point where I became loath to replace him but eventually did so in June 1988 with the arrival of Brian Richards. Tom them reverted to the role of Director, Consultant, Adviser, Critic (constructive), Lateral Thinker and Enemy of Sloppy Logic.

He has been of enormous benefit to our Company over the years of our association. He has continuously stimulated healthy discussion and review which has sometimes been uncomfortable but always constructive. His challenging style and great experience led to a respect for his views from all who have had dealings with him.

He retires in March despite my best efforts to prevent him going. I have resisted his leaving for the past 3 years but he now threatens me with physical violence. He goes with our thanks for his work and our best wishes for his future.

I would also like to record my personal indebtedness to Tom Kerr for his support, guidance, help and sense of humour which never failed him through even the most difficult times.

T. L. Grievson
CHAIRMAN

16 March, 1994

Farewell Notice. Hunting Engineering

Chapter 12

Other Activities.1970 Onwards

This chapter covers about twenty years of my other activities that do not fit easily anywhere else in this book. It was very diverse period but it was all so very stimulating.

The Royal Aeronautical Society

In 1809, Sir George Cayley wrote the first paper on Aerial Navigation laying down the principles of heavier-than-air flight. Because he was convinced that a learned society would be of the greatest benefit for progress of mechanical flight, Sir George Cayley tried three times in 1816, 1837 and in 1840 to form such a society. In 1866, the Royal Aeronautical Society was formed and is the oldest Aeronautical Society in the world. About 1924, it was proposed that the Institute of Aeronautical Engineers should amalgamate with the Society. They turned down the proposal. However, agreement was reached and in 1927 amalgamation was achieved as it was with the Helicopter Association in 1960. It became the Rotorcraft Section. In 1925, amidst great conflict and emotion, the Royal Aeronautical Society recognised the Wright Brothers as the first flyers. In appreciation of this recognition, the first aeroplane to fly was sent to the Science Museum in London.

By 1939 there were some 1100 members and, on a lease from the Crown Estates Commissioners, the Society moved into its current headquarters at 4 Hamilton Place. Over the years the membership grew to over 18,000 with many more people attending branch events. In 1986, the Society of Licensed Engineers and Technicians (SLAET) was integrated into the Society. The Society is only about 1/4 of the size of the Institutes of Mechanical and Electrical Engineers and it is always a worry that their specialist sectors are as big as our Society. However, we believe that our strength lies in the multi-disciplinary nature of our work including legal and medical specialists.

I joined the Society in 1955 as an Associate Fellow. At the Society meetings, there were excellent lectures by experts working at the forefront of the technology that was advancing very rapidly. It was possible to meet and discuss key issues with the leaders in aerospace fraternity. On many occasions between forty and sixty of us travelled by bus from RAE to these meetings. I became a Fellow of the Society in 1975 and a Chartered Engineer in 1983.

Election as President

In 1976, I was nominated and elected to the Council. In 1981, I became a Vice-President and, in May 1985, Geoffrey Pardoe handed over the Presidency to me. Without doubt, the Presidential Year, which I shall describe later, was the most exciting time.

The Society is a complex organisation. It is multi-disciplinary and operates with a small permanent staff. It is a charity. About one half of its income is provided by subscriptions and about half has to be earned by organising conferences, providing facilities for meetings and special dinner functions. The conferences and Society lecture programmes were organised by the members of the Specialist Groups. The members provide the Chairman and committees and contribute greatly to the work of the Society. We also had some 31 branches in the UK, and four Divisions and three Branches overseas. The senior committees like the Capital and Resources Committee, the Membership Committee and the Medals and Awards Committee are chaired by Past Presidents, Vice- Presidents or senior members of the Council.

The Presidential Year.

One of the tasks of the new President is to give an interview to the Aerospace Magazine reveal his views and aspirations. (Aerospace, February 1986.) In my year, we were moving towards integration with the Society of Licensed Aeronautical Engineers and Technicians, SLAET. It involved complex membership issues as well as the possibility of integrating the publications and rationalising the property of the two Societies. In addition, I wanted to move the Library from what is now the Cayley Room to some rooms on a higher floor. As a bi-product, it would also provide greater protection for our collection of historic books. Overall I wanted to produce a magnificent room for formal dinners, meetings, lectures etc., refurbish the Entrance Hall, the Council Room, the Foyer and lift. It would cost about £100k to make the changes necessary. I persuaded Mr. Henry Kramer to help. He had been a great benefactor of the Society for many years, particularly in his support for the prizes for man-powered flight. He agreed to give £75k if I could find the other £25k. Although I managed to find about half by donations to the Society, I had to resort to hiring the staircase and Cayley Room to a film company for two days to make up the deficit.

I also felt that the experience of the Council members should be available to the 19 Specialist Groups by getting the Chairmen to report in person to the Council on their past activities and their plans for the future. The Council could then influence their programmes by making suggestions for new subjects or combined group activities.

The Presidential year was frantic. One is expected to visit as many breaches as possible, including those overseas. On the overseas visits, I carried with me six twenty-minute lectures so that I could put any three of them together at a moments notice to give an appropriate lecture wherever I was. Evelyn travelled with me on every trip and the airlines were wonderful in up-grading us to business or first class so that the journeys were made in great style. We received the friendliest of welcomes everywhere we went.

In London, the big occasions for the President were- the ceremony of the presentation of the Medals and Awards, the prizes for written papers and lectures, the named lectures in

London-in particular, the Wilbur and Orville Wright lecture followed by the Annual Dinner and the Annual General Meeting followed by a fork buffet for the Members attending were very important. Evelyn and I enjoyed many Annual Dinners. I also lectured in London and at a number of the Branches.

Annual Dinner. 1988. Royal Aeronautical Society

During the year, I managed to visit about 90% of the Branches in the UK. They were always well-organised and happy occasions. The welcome was always warm and friendly. If I was asked to speak, it was a real opportunity to tell everyone what we were trying to do in London and to hear their views and concerns.

Visits to the Overseas Branches

Timed to coincide with special events, they were always wonderful occasions. Unfortunately there is insufficient space to describe all of them. Apart from the visit to Dublin, Evelyn came with me on all of them. The wives, who entertained her whilst I was busy with other appointments, gave her a great welcomed to.

Visit to Dublin — Lt. Col. John Moore invited me over for the three-day celebrations of the 50th Anniversary celebrations of Aer Lingus and, in particular, for the lecture on "The History of the Airline" by its Chief Executive. Aer Lingus started in May 1936 with a capital of £100,000, a staff of 12 and a single DH Rapide. The spares were kept in biscuit tins. That aircraft had been rebuilt with considerable care and had flown. There remained much work to be done but the restoration completed at that time was a great achievement. I had an excellent two days of visits to the Irish Air Force at Baldonnel and to the sites of the history of Irish Aviation. I was surprised and very amused at the number of Irish jokes told by the Irish during my stay.

Visit To Australia and New Zealand —Australia had branches in Canberra, Sidney, Melbourne and Adelaide. In Australia, I managed to visit them all. In the gaps, I managed to visit a number of government laboratories where discussions of our activities and problems was of mutual benefit. In Australia, the major event was the Joint National Symposium held in Melbourne between the Royal Aeronautical Society and the Institute of Engineers. On Day1, the Institute of Engineers discussed "The Influence of Aviation on Engineering" and on Day 2 the RAeS discussed "The Future of Aeronautics in Australia". I gave the Summary Address.

In New Zealand, there were Branches in Auckland, Wellington, Christchurch, Palmerston North, Hamilton and Blenheim. I managed to visit the first three. It was a hectic and very enjoyable visit where we met so many old friends. I was very heart-warming to read in the Thirty Seventh Annual Report from New Zealand in December 1985, - *In July, the Division enjoyed a visit from the London Society President. Opportunities were made for as many Division members as possible to meet him and Mrs Kerr at activities organised in Christchurch, Wellington and Auckland. Such visits by London Presidents are becoming a regular feature and it is hoped that this may continue in future years. Personal contact with London strengthens the bond between the Division and parent Society and encourages a feeling of being part of a much wider organisation that supports the common aim-" the advancement of the art and science of aeronautics. "*

Visit to Pakistan—The visit to Pakistan was timed to coincide with the Royal Aeronautical Societies Third National Conference on Aeronautical Engineering. It was well described in the English newspaper, DAWN, on 25 November 1985. Over 200 engineers attended the Conference organised by both the Institute of Engineers Pakistan and the RAeS Pakistan Branch. Mr. Dennis Little, Senior V.P. Technology and New Product Development from Airbus Industrie lectured on the Airbus. Apart from the reading of the Holy Koran and Prayers, the whole conference was conducted in excellent English. There were 23 papers and I was privileged to be Chairman for one of the sessions.

On the first evening, Evelyn and I were guests at an especially arranged dinner, at which they honoured me by awarding me an Honorary Fellowship of the Institute of Engineers, Pakistan. The Scroll was presented in a decorated silver holder.

After the conference, we were able to visit the Civil Aviation Training Institute at Hyderabad. It was a fascinating car drive to Hydrabad. Halfway, we pulled off the road into a bricked-off patch of about 4 acres in the desert. On it was one small hut in which slept the guard for the plot. He was kind enough to pull out his wooden bed for us to sit on whilst we had coffee.

At the Institute, we were briefed on the air traffic and operations centres, and the electronics and electro-mechanical schools. At this time, they were particularly proud of the "Nationalisation" of the Pakistani International Airlines. In India, before partition in 1947, there were two Muslim owned air-carriers, -Oriant and Deccan Airways. They described incidents from the early days. One was when a DC-3 landed at Gilgit on its first ever flight to that place, the residents thought that the plane was some kind of animal or bird and they brought bundles of grass to feed it. At that time, the airlines were operated by foreign management, pilots and ground staff. By 1985, the Pakistani nation felt that both the country and the airline had started against the heaviest odds and had survived and flourished. They thought that PIA, operated by Pakistani nationals, would symbolise the overall development of the country and its successful acquisition of modern technology and expertise.

Visit to Zimbabwe and South Africa. April. 1986—I really looked forward to going back to Zimbabwe and South Africa. They are beautiful countries and my time there teaching flying had been so pleasant. In Harare, the President entertained us and I lectured at a Branch meeting on "Reconnaissance- Past, Present and Future". After an all-to-short visit to Victoria Falls, we were flying to Johannesburg.

In Johannesburg, I was principle guest at the Annual Dinner of the Royal Air Force Officers Club and replied to the Toast to the Guests. We then flew on to Cape Town, I lectured again on Reconnaissance. As the guest of one of the Companies, we were able to visit the Kruger Park, the University of Witwattersrand and the Council for Scientific and Industrial Research. At the University, which was very liberal in its outlook, I was briefed on the special measures being taken to provide science and engineering courses for the disadvantaged, mainly black, students.

One of the interesting events for me was to visit a micro-light factory in Lanseria, which, as it happened, was having a flying display on that day. The Thunderbird was very sturdy for a micro-light. The workshops and construction areas impressed me. I was invited to fly in it. Although I agreed, I was somewhat unhappy about flying in a display as such occasions are not the safest for passenger flying. In this case, it would be all the more tricky, because, at ground level, we were at 5600ft altitude and everything happens more quickly than at sea level when someone like me climbs on board and the payload doubles. In one of the displays, the aircraft zoomed in over the field and pulled up steeply. Near the top of the climb, the plastic fuel tank fell off the rear of the aircraft, thudded to the ground and burst with an enormous splash. The pilot, not knowing that he had a problem, continued round the circuit, landed and was taxiing to a new take off position when the engine stopped, the fuel in the system being exhausted! Strangely, the incident encouraged me somewhat. I felt that they had had their incident for the day and, as a result, I felt that it would be safer. The flight was excellent. I was able to fly it and it was a thrilling experience. I do hope that it continued to market successfully.

Visit to Singapore, Kuala Lumpur and Hong Kong—The timing of the visit to Singapore was to fit in with an International Conference at which I was presenting a paper. After the Conference, I wanted to get to Kuala Lumpur, where the activity of the branch seemed to have stopped. I managed to contact the Chairman and the Hon. Secretary who were able to reassure me that the problems were temporary and the Branch would soon be back in action.

We went on to Hong Kong, which had a thriving branch under the Presidency of Stewart John. The combination of Cathy Pacific Airline, its maintenance facilities and the HK Polytechnic in Kowloon combined to make a powerful hive of aerospace activity. In addition, Evelyn and I were able to enjoy ourselves without the pressure of giving speeches and lectures.

The Hand-over to Dr. John Fozard

The end of the Presidential year seemed to arrive in a flash. The Awards Ceremony, the AGM and whole evening went extremely well. As I was divesting the Presidential Badge of

Office onto John Fozard he asked me what I was going to do now? I was happy to reply-"I shall go to sleep for a week!!"

Other Tasks for the Society

The Grading Committee—I served on this Committee for several years, welcoming new members and upgrading current members as appropriate.

Mechanical and Structural Group —With the Institutes of Electrical Engineering and Mechanical Engineering each having memberships in the region of 80,000 compared with our 20,000, we were always worried that they could begin to encompass some of our aeronautical territory. I was asked to set up a new committee called "Mechanical and Structural Group" to initiate a programme within the Society and to collaborate with the Aeronautical Structures Group within the Mechanicals. I ran it for about a year and then handed over to Frank Vann of BAe. It was successful and the collaboration worked.

The Medals and Awards Committee—It was a great honour to follow Sir Charles Pringle as Chairman of this Committee. Its task is to make recommendations to the Council on the winners of Medals, Awards and Prizes each year. This task involved two three-month periods of intense activity, one before the Wilbur and Orville Wright Memorial Lecture in December, when the Medals and Awards for the year are presented, and one before the AGM in May, when the prizes for Papers in the Aeronautical Society Journals are presented. In 1990,there were some 55 RAeS Awards and Prizes. The nominations for the prizes at the Universities were made by the appropriate Professors and by the Course Commanders at RAF Halton and Cosford. Overall, I was always left with the feeling that, despite the number of awards and prizes, there were many worthy candidates who did not receive one. We also made recommendations to other bodies who made awards in the aerospace field, such as the Guild of Air Pilots and Navigators, the Air League, etc.

We attracted many applications for financial awards to assist with new innovations in the aerospace field. The Handley Page Award enabled grants of up to £5000 to be available each year for original work leading to the advancement and progress in the art and science of aeronautics directed towards the safety of those who worked with or travel in aircraft. One application that gave me particular pleasure was received, in 1992, from James Labouchure seeking funding support to establish a new design of seaplane hull without using the conventional step.

A revolution in air-land-sea transport, recreation and convenience

The Centaur Flying Model.

We recommended a series of tests on different hull shapes to measure the forces on the hull through the un-stick speed range, using an apparatus fitted to a speedboat and were able to support him with appropriate funds. With the success of these trials, we were able to assist him with the building of a powered model of the Centaur, to demonstrate the low hull shock loading, the clean aerodynamics and the low hydrodynamic drag during take-off. It also provided, via the low shock loading, the larger wave handling capability some 1.8 times better than conventional hulls and the design allowed good waterside accessibility.

It is a pleasure to hear that the first aircraft is being built. The Society wishes it every success.

The Francois-Xavier Bagnoud Prize. — In 1992, as Chairman of the RAeS. Medals and Awards Committee, I was invited to be a member of the first awards committee for an American Aerospace Prize, the Francois-Xavier Bagnoud for $250k, awarded once every two years.

A citizen of Switzerland, Francois-Xavier went to the United States for university training and graduated from the University of Michigan in 1983 with a degree in Aerospace Engineering. An avid pilot, he became, at 23 years of age, the youngest professional IFR pilot in Europe of both aeroplanes and helicopters. In 1986, tragically, he flew a fatal helicopter mission in Mali, West Africa. In 1989, his mother and father, Countess Albina du Boisrouvray and Bruno Bagnoud with friends founded an Association " to promote human development projects that embody and continue her son's values and those things dear to his heart." In one of the Aerospace projects undertaken by the Association, a prize was to be awarded for outstanding accomplishments in the aerospace field, The primary consideration was to be given to innovation and accomplishment in aerospace engineering, science, and medicine resulting in important benefits and significant advancements to the well-being of humanity. The prize may be awarded to an individual or group of contributing individuals for a specific achievement or body of work extending over a period of years. Candidates must be living persons at the time of selection for the prize.

The selection committee consisted of a Chairman and twelve members chosen by the FXB Board. Members were expected to attend one meeting in Ann Arbor, Michigan, USA, when the final selection was to be made. In that year, there were thirty-two nominations of which 17 were American, 3 British, 4 Russian, 3 Chinese, 3 German and one each for India and Sweden. The Chairman asked us to select the top five and rank them in order. It was interesting that a clear preference for the same three were included in the selections by each member of the panel.

Dr. William H. Pickering was chosen. Part of his citation read," More than any other individual, Bill Pickering was responsible for America's success in exploring the planets, — an endeavour that demanded vision, courage, dedication, expertise and the ability to inspire two generations of scientists and engineers at the Jet Propulsion Laboratory (JPL)." When presented with his Prize by the Chairman of the Board, he presented a paper entitled, "Some Reflections on Space Research-The Challenges and the Triumphs."

We also attended the dedication of the Francois-Xavier Bagnoud Building that had been supported by a large number of contributors. It was a splendid modern structure containing both classrooms and laboratories. In all, I spent three days on the University of Michigan campus. It was delightful and I had the pleasure of escorting German I. Zagainov, Head of the Central Aerodynamics Institute, Russia, who was accommodated in a room close to mine.

The Chairmanship of the Bedford Branch of the R.Ae.S. —I was invited to become Chairman in 1992 for one year. In addition to the normal duties of a Chairman, he is expected to deliver a lecture during the year. Having worked at RAE Bedford for five years, been responsible for it, as Director RAE, for four years and a Director at Hunting Engineering located just south of Bedford for eight years, I felt that I was able to lecture on the contribution made by Bedfordshire to the Advances in Aerospace. That year took me back to RAE Bedford and Cranfield numerous times and it was always a pleasure meeting old friends again.

The Visit to Chinese Society of Aeronautics and Astronautics CSAA—The Chinese first approached the RAeS during 1985. After discussions with them, we received an official invitation dated 24th December 1986 for a party of six RAeS members and their wives to visit China during April 1987. Each member of the team was asked to suggest titles of lectures that they might give in China from which the Chinese could make their choice. The RAeS team included John Fozard, President, Prof. John Stollery, Geoff Howell, P. R. Sampson, Alan Newton and myself. The wives of the first three, Gloria, Jean and Margaret respectively, and Evelyn went with the party.

The topics that we offered for lectures were:-
JF -Present and Future V/STOL Tactical Aircraft, and Advanced Jet Trainers and Light Attack Aircraft,
JS- Wing Sections for Remotely Piloted Aircraft plus four others,
TK-. Aerospace Engineering into the next Century,
GH- Civil Aviation Safety,
PS-Mechanical Engineering Aspects of Airborne Radar Systems and
AN- Low Specific Thrust Engines.

The team plus wives left for Hong Kong on 12th. April.87 aboard a Cathy Pacific 747 all upgraded to First Class. We then flew on to Beijing. The next day, we were met by three Professors, two of whom had been to Cranfield and Prof Shen Yuan whom I had met on the previous visit to China. We split into three pairs for our lectures, in a session of three hours. The lectures were repeated in several places in China.

The next day we visited the Institute of Materials in the morning and then had a meeting with the Minister of Aviation in the afternoon. We had a long discussion on the future of the latest technology and the next generation of supersonic aircraft. The Minister was also very keen on the next generation of supersonic transport and, in particular, Hotol. That evening he hosted us all to a splendid dinner and the next day sent us off to visit the Ming Tombs. Since Evelyn and I had last been there, they had moved many of the displays of the contents of the tombs to other museums. The next day, we visited the Great Wall and the Forbidden City.

In Xi'an, we were accommodated in the guesthouse of the North-Western Poly-Technical University. We visited the Terracotta Warriors, the Hot Springs and Banpo Village, some 6000 years old. The next day we all lectured at the NPU. The following day, we were due to fly to Nanking but the flight was delayed all day. We eventually arrived in in the University at 7.00pm and were put into student accommodation. Not only was there no hot water but also there was no food available as the university facilities were closed. With our escorting Chinese representatives, we organised some transport to a large western style hotel to have dinner. The contrast was almost unbelievable. We sat at two tables, seven on each, and enjoyed a splendid dinner in an air-conditioned dining room. Towards the end of the dinner, John wandered over to me and said, " I'll pay for the other table, Tom would you pay for this one?" It was all worth it. That night, I discovered that there was a big hole in the mattress of my bed, in a position where my posterior fell through. Luckily, I was able to change beds with Evelyn who, being shorter, was able to sleep with her legs across the hole.

The next day, we toured the university facilities. Later, several of us lectured again, whilst the others went on a sightseeing tour. We saw the longest combined road/rail bridge over the Yangtze. Most of the hosts, having studied in Britain, spoke good English.

The weekend was spent at Hangchow, a delightful holiday area with numerous lakes. We travelled by train and arrived there without our luggage including our soap bags. Despite the beauty of our surroundings, it was hot, there was torrential rain and, to say the least, the ladies were getting desperate for a change of clothing. The luggage had not arrived by the time we were due to leave for Shanghai on an express train on the second evening. Whilst on the platform, by chance we saw the luggage in a heap further down the platform. Our guides assured us that it would follow us immediately on a slow train. That wasn't good enough. The ladies led the charge and the baggage was heaped onto the express in our compartment.

In Shanghai, we all lectured again. Then we spent two days sightseeing including a trip on the Huangpu River to where it joins the Yangtse. At that point, it is so wide that it is quite impossible to see the other bank. We also visited the Acrobats Theatre.

We flew to Hong Kong to complete a very stimulating trip. For Evelyn and I, it was very interesting to see the changes since our last visit. The girls had become much more feminine and colourful. There was a much freer atmosphere and there were significant improvements in the whole society. It was also a great pleasure for us to meet again Professor Kang Yi and Miss Wong when we were in Beijing.

The Parliamentary and Scientific Committee—I represented the RAeS on this Committee for more than five years. The meetings were held in the House of Commons approximately eight times per year when there were opportunities to hear experts in the latest scientific problems or advances and to hear the political reaction to them. I was surprised to find the depth of the emotion involved in some of the subjects and I was very interested in topics that were completely outside the aerospace field. Two areas of particular interest were Public Health and Global Warming. A group of us, lead by Lord Flowers, visited one of the 50 clinical and environmental laboratories of the Public Health Laboratory Service. We were briefed on the key role of the laboratory based communicable disease surveillance function and given briefings on legionnaires disease, hepatitis, meningitis, measles and HIV. In another group, I visited the British Antarctic Survey Laboratories and was briefed on the current position on many studies. To quote, *"Science bites" from the Antarctic seem to leave a deep impression on the media and the public alike. But like icebergs, nine-tenths of the science remains below the surface, unrecognised and unreported.*

It seems to me that the problem of global warming was not being examined broadly enough. The world has been through a number of ice ages before mankind could possibly have influenced it. We have no explanation of the ice ages that have been and gone. The Antarctic ice core studies have shown that the most common greenhouse gases have increased each time the global climate has warmed and decreased when it has cooled but humans made no contribution to those events. The British Arctic Survey participation in the multi-national EPICA project could retrieve 250,000 years of climatic history. In time we may begin to appreciate some of the main influences on the problem.

Brooklands Aviation.

Alan Curtis was the first of many who were interested in introducing Business Aviation into Farnborough airfield. His flamboyant enthusiasm for all he did and the financial backing he was able to obtain made him a very significant player in the exciting innovations of life. He was Chairman of Lotus and Aston Martin and owned an airfield. Our contact continued for a number of years. On one occasion, he invited me to be a referee in the Nouveaux Beaujolais race from France to London. That event was full of exciting moments.

We flew to France to pick up the Beaujolais. After having dinner, the wine could be collected at midnight and, from that moment the race to the Savoy Hotel in London was on. Stirling Moss was one of the contestants. Despite a number of incidents, *(broken-down cars seemed to be scattered across the south of England)* the wine arrived. The following day a major charity event was held and a considerable sum was raised.

Next, members of the Airborne Division, notionally having flown from France, parachuted into the Thames to the deliver the wine. The parachutists were rescued by boats on the Thames, transferred to the riverbank from whence they raced to the winning post. The parachutists were forbidden to carry the wine during the jumps. It was hidden in the boats on the Thames for their collection. The competition was so intense that some of the parachutists did carry wine during their jump. In one case, the snatch of the parachute as it opened was so fierce that the bottle escaped into freefall over London. It fell into Covent Garden, close enough to a lady to splatter her with red wine. With compensation for the damage this very worrying incident was closed.

The Optica, a unique observation plane, was designed and built by John Edgley in his back yard. With financial backing of some £9m, he developed production facilities at Old Sarum. In May 1985, whilst on exercise with the Hampshire police, the first aircraft crashed killing the pilot and passenger. Though the aircraft was subsequently cleared of blame, four months later the company was in receivership and all but 50 employees made redundant. It was rescued but by autumn 1985, only one aircraft had been sold and the business was again insolvent.

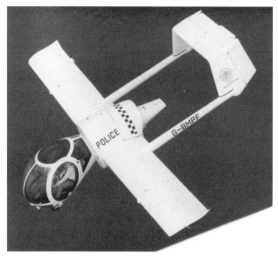

The Optica Aircraft

On 1st. January 1986, Alan Haikney saved the company from a third receivership by acquiring 50%, injecting the necessary funds and assuming control. Construction of the aircraft continued. The decision was taken to widen the scope of the company by establishing profit centres for design and manufacture providing subcontract work for the aerospace industry. By then five aircraft had been built but none sold. On 17th Jan 87, there was a fire in the hanger that was destroyed with five Optica and other aircraft. Even the aircraft parked outside had been deliberately set on fire and were damaged beyond repair. In effect, the company was bankrupt. The police believed that arson was involved and there was no incentive for the insurance money to be paid. After clearance by the police, payment by the insurance company and again an input from a friendly investor saved the company.

Later that year, Brooklands Aerospace Group was formed with four subsidiaries, Brooklands Aero Manufacturing to provide subcontract manufacturing to the major aerospace companies, Brooklands Aero-Technology to provide aerospace design skills for the industry, Brooklands Airport Support to provide an aircraft hover recovery system, mobile taildocks for jumbo jets, aircraft steps etc, and Brooklands Aircraft Company to market and support the Optica aircraft. There was also Brooklands Property Ltd to pursue the development of a site of some 160 acres. The eventual development would encompass another 80 acres leaving the airfield as a major amenity.

In December 87, the Hampshire police became the first purchaser of an Optica Scout. It was a greatly enhanced version of the original Optica, having undergone some 44 changes to improve the operating performance etc. By then Alan Curtis had joined the Board and Mathew Hudson joined soon afterwards.

In January 1989, Alan Haikney sold his interest in Brooklands Aerospace and Alan Curtis took over as Chairman. He completed a financial restructuring in which a Mercantile Investment Trust took a major equity interest. The rest of the shares were owned by a group of investors including the Directors and Management. At that time, all looked well for the company. The American Federal Aviation Authority was close to giving the Optica its certification, there seemed excellent prospects for sales in the USA and to Middle and Far East countries, Australia and Africa. The four subsidiaries seemed to be doing well and there were other interests in aircraft like the Bulldog, the Chipmonk, Fieldmaster etc.

It was in January 1989 that Alan Curtis invited me to become a non-executive Director. On the Board, the combination of himself and Mathew Hudson was an extremely powerful sales and marketing force. Perhaps he thought that some technical strengthening might be desirable.

It was also evident that, to have any hope of trading profitably, the manufacture of the Optica and any other small aircraft would have to be done outside the UK. We needed to get the built price down to £100k to be able to sell at £135k to cover all other expenses and make a profit. Considerable effort was put into getting agreements with RO FABRIKA "UTVA" and JUNGOMETAL of Yugoslavia to manufacture the Optica, Venture SHB 1 and Bulldog aircraft. The final assembly would be in the UK.

An Optica had been shipped by sea to the USA to be the focus of a major sales campaign. Unfortunately, the ship ran into a storm and the aircraft was soaked in seawater and was a write-off. The insurance company took a long time to agree the claim. In addition, another aircraft was damaged in Spain. We were beginning to loose our credibility.

In May 89, the FAA called for further studies. Firstly, on the requirement for an emergency fuel dump system to allow the maximum all-up-weight to be reduced in the event of an emergency and, secondly, on the fatigue life of the aircraft. The fatigue analysis, carried out as part of FAA certification, indicated that the fatigue life of the cabin attachment lugs was only 9000 hours and not 20,000 hours as calculated by Edgley. The fatigue life depends upon the type of flying that the aircraft does and the resulting load spectra that it experiences. The FAA also produced some new data gathered from flights to check the turbulence at or near forest fires, pipeline patrols over level and mountainous terrain i.e. the typical roles that Optica would be expected to perform. Unfortunately the data showed loading spectra that were far more severe those that used in previous calculations. New calculations indicated a "life" of approximately 1000hours.The "home build" joints of both the wing lugs and the cabin lugs needed redesign if a fatigue life of 20,000 hours was to be offered. The aircraft would also require fatigue testing on the ground.

By July 89, the company cash flow position had become critical. No aircraft had been sold and there was a shortfall in the sales of all other subsidiaries. Though the major shareholders were prepared to inject further funds by December 89, creditors were calling in their loans and the Receiver was called in. The Receiver offered the company for sale. There was some interest and the prospects of a sale were high until the very last moment. Eventually Lovaux, part of FLS Aerospace, bought it and announced that Optica development would continue. I have not heard of the Optica since that time!

The Guild of Air Pilots and Air Navigators. GAPAN.

With the strong support of Freddie Stringer, Air Vice-Marshal Michael Adams and others, my application to become an Upper Freeman of the Guild was approved in 1988. I enjoyed taking part in the activities of the Guild particularly the visits, lectures and dinners, which were splendid occasions. By 1996, my back problems had made it so difficult to travel to London, that I felt it right to offer my resignation. On the day it was discussed in Council, my old friend Air Chief Marshal Sir Neil Wheeler was in attendance. He said that they should endeavour to persuade me to change my mind. He argued that the Guild had not offered me the Livery and it should. Thus followed a series of discussions during which the promotion to Liveryman was recommended. I agreed to continue and hope that my back would have improved.

To become a Liveryman, I first had to become a Freeman of the City of London. The Freedom of the City of London is an ancient tradition. It is not an award or an honour unless conferred as the rarely granted Honorary Freedom. The Corporation of London regulate the Livery Companies to a degree and demands that all liverymen be freemen of the City as a prerequisite to trade, in exchange for certain valuable trading rights and privileges.

Today, about half the applicants for the freedom are "presented" by a City of London Livery Company. In those circumstances, the copy of the Freedom is written with the full names of the Freeman and the Livery Company, with *R.Co.Ald.* (Redemption, by Order of the Court of Aldermen). The short but solemn freedom admission ceremonies take place in the Chamberlain's Court Room in the Guildhall. Generally each ceremony is conducted individually and the candidate admitted by the Court of Alderman. I was appointed a Liveryman of GAPAN on 8th August 1996 and welcomed by the Duke of Edinburgh, the Grand Master, when he attended the Court to celebrate his 75th birthday at the Court Dinner on 2nd October.

The Society of British Aircraft Companies.

SBAC—I served on the Council and chaired the Guided Weapon Sectoral Group for two years. Within the GW group was a Standing Committee on Guided Weapon Research and a Standing Committee on Guided Weapons Range Facilities.

In the same year, we tried to bring together a view of the implications of current UK and European policies on the guided weapons industry, "Guided Weapons into the Next Century. The Implications of Current Policies to the G.W. Industry in the U.K." The guided weapon industry had grown dramatically since the latter stages of WW11. In the 1960's and 1970's as costs rose to match the capabilities required, collaborative developments were initiated usually on a bilateral or trilateral basis. Since that time, the costs of research, development and production had continued to escalate with the requirements for increased operational capabilities. International collaboration had become the rule rather than the exception.

The Independent European Programme Group (IEPG) commissioned a European defence industry study "Towards a Stronger Europe" published in 1986. The report recommended measures to strengthen the industry in Europe. Greater co-operation in R&D

was foremost amongst them. Two very specific recommendations were - the harmonisation of military requirements and procurement timescales and the preparation of a document summarising the National Objectives and Intentions for International Defence Equipment Collaboration. It was a forlorn hope. It was sad because our study of the possible future GW programmes indicated that most, if not all, would be collaborative.

Europe was, and still is, a complicated place from the point of view of Government Treaty Organisations. It includes the then current members of NATO, WEU (The Western European Union), Council of Europe, the EC(Economic Community) and the IEPG, the Independent European Programme Group. In order to interface as a body with these organisations, the European Industry had formed the Association Europeene Des Constructeurs De Materiel Aerospatial (AECMA.) Its Terms of Reference were- *to ease the European industries use of major existing experimental facilities and to promote the creation of new installations, not yet available on a European basis, to meet industry's requirements.* SBAC appointed me as their representative and proposed me to chair the Committee. We tried to get at least one representative from each country and were expected to meet three times a year. We were tasked with reporting to the Technical and Industrial Commission.

Our first major task was to monitor, review and analyse the situation of the existing major experimental facilities available to the European Aerospace Industry and their test techniques, and compare them with those available in the USA. The bulk of this task was completed and published in two years against a background of reduced defence spending and the rationalisation of a decreasing aerospace industry, and the difficulties of finding appropriate support from the companies.

I think that the development problems were summed up by Alan Clarke, Minister of Procurement during his address at a Bow Group Conference when he said - A request for approval for a new Army vehicle had been put to him. He decided that it could be a candidate for a collaborative project and he would ring his European opposite numbers and seek a partner. Much to his dismay, they all turned him down. When he came to sign the approval he realised that the order was the same size as that for the Californian National Guard. The separate European countries, even the larger ones, do not buy enough units to go it alone.

The MoD procurement was the dominant influence on the size and shape of the UK industry. Competition had introduced a new dimension into the procurement equation. The UK industry was not only competing between UK companies but also with overseas companies whose governments had taken a different view of the support to industry. There was a further complication in that in the UK, MoD was the main procurer/buyer and thus had considerable power, whereas the DTI had the responsibility for industry and had little power to influence anything.

The Federation of Electronic Industries, FEI.

I served as the representative of Hunting Engineering on the Council of the Federation of Electronic Industry for five or more years and on their Strategic Policy Committee. It represented some 236 member-companies with an annual turnover of £65bn. They were major providers of capital equipment, systems and components for military communications, information technology, defence electronics, aerospace and maritime sectors.

One of the interesting phases was the contribution of their views to the House of Lord's Select Committee on Science and Technology and, in particular in 1993, the evidence to Sub-Committee 11 on the Defence Research Agency. As the companies were dealing with front-edge technology, the member's lifeblood was research and thus the health and efficiency of the Defence Research Agency was of prime importance to them.

The DRA had operated as a Trading Fund since April 1991. Over many years, a good working relationship existed between the Research Establishments and Industry. There was considerable concern about how the new relationships with the MoD, the DTI, the DRA and industry would be formed and the impact on the future of civil and defence research, particularly dual-use research. The reorganisation of the Defence Research Establishments into the DRA, particularly the financial management systems as well as the Research Package Management or the various research and project areas was a major unknown. Industry was also worried about the potential competition between the DRA and industry and how the Intellectual Property Rights would be handled?

In the weeks ahead of our date with the select Committee, the FEI had prepared detailed answers to twenty three questions and then answers to a further nine supplementary questions. On the day, Eric Brydon, Chief Engineer GEC-Marconi, John MacNaughton, Director Electronic Data Systems and I appeared before the Committee, Sir Ronald Mason, previously Chief Scientific Adviser to the Ministry of Defence and previous Chairman of Hunting Engineering, was due to give evidence ahead of us. We were able to sit and listen to him. I was particularly interested in what he had to say as one of the things we had tried, during his period as Chief Scientific Adviser, was to set up an Defence Technology Enterprises Ltd, whose specific task was to sell or lease to industry, some of the dual-use technologies from the Research Establishments. That company was not very successful. Today the company exists only to collect a few royalty payments. We were now in a situation where the DRA was expected to sell the bulk of its technology to everyone. There was also a concern whether other countries, particularly the USA, would see the DRA as an Institution of the State, particularly from the point of view of Nuclear Systems, Electronic Warfare and Cryptography. In fact as there were so many questions to which the answer was not available, it was difficult to see the future at all clearly.

In the verbal evidence, we tended to emphasise the problems that were becoming visible, i.e. the worry that with shrinking research budgets and the commercial imperatives placed on the organisation, that the DRA would attract more research to itself rather than industry. In particular, it might want to do more short-term research because of the immediate commercial value rather than the longer term and systems work that had long been the Research Establishment's forte. Industry felt that its strengths were in assessing the commercial potential of short-term research and it had the ability to market the products successfully. The visibility of the research programme was being reduced. Software work might be pulled back into the new DRA Software Engineering Centre. Industry also had concerns that the Trials and Evaluation Agency would be absorbed back into DRA. Industry makes use of these facilities and has to pay a fee. However, when competing against the DRA for a contract, industry felt at a distinct disadvantage because the DRA seemed to be using "opportunity costs" to give it a price advantage for some tasks.

I think that the presentation of our evidence on the day went well and many of our fears have been confirmed. It was also a particular pleasure for me to meet again Lord David Craig and Professor Hartley.

"Farnborough, Above & Beyond". A Meridian ITV Production.

In 2000 Onyx Productions, funded by B.Ae., DERA, and others produced six thirty-minute programmes for Meridian TV covering the whole history of RAE from the first few years of the twentieth century. The programme presented many retired members of RAE telling the story of their role and how the work was done as well as archive film footage, capturing the pioneering spirit of the time. I was filmed for about two hours talking about the main areas of activity. It covered Aerodynamics, Structures, Flight Systems, Weapons, Space and Nuclear Weapons. Only about six minutes of that filming appeared in the broadcast programmes and nothing about Nuclear Weapons. The presentation of a century of aviation history at Farnborough helped to tell an exciting story that may never be fully told.

It was a wonderful life!!